THE SCIENCE OF INSTREAM FLOWS

A Review of the Texas Instream Flow Program

Committee on Review of Methods for Establishing
Instream Flows for Texas Rivers

Water Science and Technology Board

Division on Earth and Life Studies

NATIONAL RESEARCH COUNCIL
OF THE NATIONAL ACADEMIES

THE NATIONAL ACADEMIES PRESS
Washington, D.C.
www.nap.edu

THE NATIONAL ACADEMIES PRESS 500 Fifth Street, N.W. Washington, DC 20001

Support for this study was provided by the Texas Water Development Board under Contract No. SLOC 2003-483-494. Any opinions, findings, conclusions, or recommendations expressed in this publication are those of the author and do not necessarily reflect the views of the sponsor.

International Standard Book Number 0-309-09566-2

Additional copies of this report are available from the National Academies Press, 500 5th Street, N.W., Lockbox 285, Washington, DC 20055; (800) 624-6242 or (202) 334-3313 (in the Washington metropolitan area); Internet, http://www.nap.edu

Printed in the United States of America.

THE NATIONAL ACADEMIES

Advisers to the Nation on Science, Engineering, and Medicine

The **National Academy of Sciences** is a private, nonprofit, self-perpetuating society of distinguished scholars engaged in scientific and engineering research, dedicated to the furtherance of science and technology and to their use for the general welfare. Upon the authority of the charter granted to it by the Congress in 1863, the Academy has a mandate that requires it to advise the federal government on scientific and technical matters. Dr. Bruce M. Alberts is president of the National Academy of Sciences.

The **National Academy of Engineering** was established in 1964, under the charter of the National Academy of Sciences, as a parallel organization of outstanding engineers. It is autonomous in its administration and in the selection of its members, sharing with the National Academy of Sciences the responsibility for advising the federal government. The National Academy of Engineering also sponsors engineering programs aimed at meeting national needs, encourages education and research, and recognizes the superior achievements of engineers. Dr. Wm. A. Wulf is president of the National Academy of Engineering.

The **Institute of Medicine** was established in 1970 by the National Academy of Sciences to secure the services of eminent members of appropriate professions in the examination of policy matters pertaining to the health of the public. The Institute acts under the responsibility given to the National Academy of Sciences by its congressional charter to be an adviser to the federal government and, upon its own initiative, to identify issues of medical care, research, and education. Dr. Harvey V. Fineberg is president of the Institute of Medicine.

The **National Research Council** was organized by the National Academy of Sciences in 1916 to associate the broad community of science and technology with the Academy's purposes of furthering knowledge and advising the federal government. Functioning in accordance with general policies determined by the Academy, the Council has become the principal operating agency of both the National Academy of Sciences and the National Academy of Engineering in providing services to the government, the public, and the scientific and engineering communities. The Council is administered jointly by both Academies and the Institute of Medicine. Dr. Bruce M. Alberts and Dr. Wm. A. Wulf are chair and vice chair, respectively, of the National Research Council.

www.national-academies.org

COMMITTEE ON REVIEW OF METHODS FOR ESTABLISHING INSTREAM FLOWS FOR TEXAS RIVERS*

GAIL E. MALLARD, *Chair*, U.S. Geological Survey, Westerly, Rhode Island
KENNETH L. DICKSON, University of North Texas, Denton
THOMAS B. HARDY, Utah State University, Logan
CLARK HUBBS, University of Texas, Austin
DAVID R. MAIDMENT, University of Texas, Austin
JAMES B. MARTIN, Western Resources Advocates, Boulder, Colorado
PATRICIA F. MCDOWELL, University of Oregon, Eugene
BRIAN D. RICHTER, The Nature Conservancy, Charlottesville, Virginia
GREGORY V. WILKERSON, University of Wyoming, Laramie
KIRK O. WINEMILLER, Texas A&M University, College Station
DAVID A. WOOLHISER, U.S. Department of Agriculture, Agricultural Research Service (Retired), Fort Collins, Colorado

NRC Staff

LAUREN E. ALEXANDER, Study Director
DOROTHY K. WEIR, Senior Program Assistant

---

* The activities of this committee were overseen and supported by the NRC's Water Science and Technology Board (see Appendix B for listing). Biographical information on committee members is contained in Appendix C.

Preface

Instream flow science is an evolving field that brings together aspects of hydrology and hydraulics, biology, physical processes and geomorphology, and water quality. Instream flow programs are being developed to answer the often politically-charged question, "how much water should be in the river?" To balance ecologic and economic uses of water, instream flow programs rely on scientific input within a legal, social, and policy context.

The act of combining science and policy into a coherent, operational instream flow program is a challenging task. Across the United States, municipalities, counties, and states grapple with issues of ensuring adequate water in times of high demand and low supply. Texas has developed a prospective instream flow program to address these challenges. With its range of river and ecosystem conditions, growing population, high demands on water and episodic water scarcity, Texas in many ways is a microcosm of instream flow challenges across the United States, and its instream flow program may serve as a template for other jurisdictions.

Our NRC committee was charged to evaluate the Texas Instream Flow program as described in the Texas Instream Flow Programmatic Work Plan (PWP) and the Technical Overview Document (TOD). This report is the result of the National Research Council's (NRC) Committee on Review of Methods for Establishing Instream Flows for Texas Rivers review of the Texas instream flow program. We were asked to comment on a technical work that already had been prepared by scientists and engineers in the state agencies. (See *http://www.twdb.state.tx.us* for the full text of the documents). In addressing our charge, the committee resisted the temptation to produce an overly prescriptive report, as it was not our assignment to (re)design the Texas instream flow program or to write an instruction manual of how to conduct an instream flow study. A prescriptive approach, which could involve detailed recommendations about techniques and methods or even a rewrite of the technical documents, would not have been appropriate. Furthermore, an overhaul of these documents did not prove necessary because the state agencies set forth a proposal with most of the important elements of a comprehensive instream flow program. The committee's review, instead, identifies missing parts and recommends bolstering the skeletal pieces of Texas' proposed program.

In preparing this report, the committee benefited greatly from our conversations with Texas State agency personnel who helped us understand the background for the Texas instream flow program. Without exception, they were open and responsive to our queries about Texas water resources and the multiple demands on water in the state. State agency personnel also helped us gain a better understanding of how the PWP and the TOD were prepared, including the difficulties of producing a plan by three agencies with three different missions.

The committee felt it would be a disservice to the Texas state agencies if we neglected to comment on the need for clear and measurable goals and a discussion of implementation. Clear, measurable goals and pragmatic ways to achieve those goals are critical to a successful instream flow program. Goal setting is the realm of policy makers, stakeholders, and other decision makers, but scientists have an important role in setting goals of an instream flow program as well.

Implementation of instream flow recommendations in Texas occurs in a complex setting where there are multiple and competing needs for water. Means to implement instream flow recommendations are necessary to prevent wasted time and resources of conducting technical evaluations of hydrology, biology, physical processes, and water quality. Oftentimes, programmatic aspects of implementation are not directly tied to the technical pieces of an instream flow recommendation. However, programmatic aspects establish important legal and pragmatic boundaries for the instream flow scientific studies and, thus, are discussed in this report.

A variety of water resources stakeholders in Texas including river basin authorities, municipal agencies, the academic community, non-governmental organizations, agricultural interests, and other citizen groups helped us understand the importance of stakeholder involvement in setting instream flow goals and establishing instream flow recommendations. The committee held three of its four meetings in Texas. During the open sessions of these meetings we heard public comment on the state's instream flow program; we learned that the public holds strong conviction on river management priorities. In all, the public participation experience of this committee in Texas, in keeping with experience in other parts of the country, underscored the import of stakeholder participation and a fair, open, transparent process for determining instream flow in Texas.

Because instream flow science is new and still evolving, we provide a short tutorial (Chapter 3) that reflects the most current thinking on the subject. Texas' prospective and systematic plan for its instream flow program gives the state an opportunity to establish a benchmark instream flow program and make significant contributions to the science. Our committee hopes that the findings and recommendations contained in this report will help the state and others realize this advancement.

We have many people to thank for their help over the course of this project and in the preparation of this report. The Texas agency personnel were incredibly supportive of our committee and its progress towards report completion. They were particularly instrumental in organizing and leading field trips for the committee to see and experience the beauty and complexities of Texas river ecosystems. We express appreciation to Barney Austin and Bill Mullican, Texas Water Development Board; Kevin Mayes, Texas Parks and Wildlife Department, and Doyle Mosier, Texas Commission on Environmental Quality; and the staff of Texas State University at San Marcos, Joanna Curran, Marshall Jennings, and Andrew Sansom. We also thank panel participants Mary Kelly, Richard Kiesling, Barbara Nickerson, Dianne Wassenich, and William West, Jr.; and other guest presenters Todd Chenoweth, Kevin Craig, Mark Fisher, Ronald Gertson, Myron Hess, Kenneth Kramer, Ren Lohoefener, Greg Rothe, and Kenny Saunders. The report and the study process would not have been possible without the hard work of NRC study director Lauren Alexander and project assistant Dorothy Weir. Finally, I would like to recognize my fellow committee members for their long hours and dedication to advancing the science and art of instream flows in Texas.

This report was reviewed in draft form by individuals chosen for their diverse perspectives and technical expertise in accordance with the procedures approved by the NRC's Report Review Committee. The purpose of this independent review is to provide candid and critical comments that will assist the institution in making its published report as sound as possible and to ensure the report meets institutional standards for objectivity, evidence, and responsiveness to the study charge. The review comments and draft manuscript remain confidential to protect the integrity of the deliberative process. We wish to thank the following individuals for their review of this report: David Ford, David Ford Consulting Engineers, Inc.; Jim Geringer, former Governor of Wyoming; Douglas James, National Science Foundation; Ronald Kaiser, Texas A&M University; Robert Milhous, U.S. Geological Survey; Bruce Rhoads, University of Illinois; Clair Stalnaker, U.S. Geological Survey (retired); and Peter Whiting, Case Western Reserve University.

Although the reviewers listed above have provided many constructive comments and suggestions, they were not asked to endorse the conclusions or recommendations, nor did they see the final draft of the report before its release. The review of this report was overseen by Kenneth Potter, University of Wisconsin. Appointed by the National Research Council, he was responsible for making certain that an independent examination of the report was carefully carried out in accordance with the institutional procedures and that all review comments were carefully considered. Responsibility for the final content of this report rests entirely with the authoring committee and the institution.

Gail E. Mallard, *Chair*

Contents

Executive Summary

Texas has more than 190,000 miles of relatively flat, warm-water streams and rivers that sustain important habitat for some 250 species of fish and provide water resources for 20 million people. Rivers in Texas exhibit considerable biotic variability that reflects the state's varying climate, geology and soils, and topography. The patterns of water availability and water use across the state are not always coincident, leading to episodic water shortages.

Variable river flow conditions in Texas combined with rapid population growth and competing demands from irrigators, recreationalists, conservationists, and municipalities spurred the creation of a statewide instream flow program in 2001. Texas Senate Bill 2 (2001) instructed three state agencies—the Texas Water Development Board (TWDB), the Texas Parks and Wildlife Department (TPWD), and the Texas Commission on Environmental Quality (TCEQ)—to develop a state program for instream flows to support a "sound ecological environment" on priority rivers by the end of 2010. In response, the agencies drafted a proposed instream flow program that is described in two documents: the Programmatic Work Plan (PWP; TPWD, TCEQ, and TWDB, 2002) and Technical Overview Document (TOD; TPWD, TCEQ, and TWDB, 2003). The PWP outlines the programmatic elements of the instream flow initiative, and the TOD details scientific and engineering methodologies for data collection and analysis. The agencies arranged for the National Research Council (NRC) to evaluate the Texas instream flow program, including the PWP and the methodologies in the TOD and other supporting documents. The NRC appointed a committee to carry out this assignment. Its statement of task is given in Box ES-1.

INSTREAM FLOW SCIENCE AND PROGRAMS

The field of instream flow science has grown rapidly over the past few decades, with many research studies and initiatives in progress in the United States and around the world. Still, instream flow science and practice are

BOX ES-1
Statement of Task for Texas Instream Flows

The committee will appraise the scientific and engineering methods used to help establish instream flow recommendations in Texas rivers, and focus on the soundness and adequacy of the Programmatic Work Plan for developing instream flow studies developed by the TWDB, TCEQ, and TPWD. Specifically, the NRC committee will:

1. Evaluate the key documents that explain these scientific and engineering methods and their applications in setting instream flow recommendations. These documents are a) the 2002 Programmatic Work Plan, and b) a supplementary technical volume that describes these methods in greater detail.
2. Review and provide advice on several scientific and technical matters relevant to instream flow studies and recommendations, including:
    a. appropriate spatial scales of analyses in hydrologic and related models;
    b. use of habitat-flow relations in setting instream flow requirements;
    c. use of landscape ecology metrics in setting instream flow requirements;
    d. range of biophysical model parameters employed in the Texas State TMDL program;
    e. applicability of water quality models used in the Texas State TMDL program to instream flow studies.
3. Evaluate findings and recommendations of Tasks 1 and 2 for consistency with the requirements of Texas law for the study of instream flows

relatively new, and basic premises of this field continue to evolve. How flow regimes influence the structure of aquatic and riparian ecosystems is largely unknown, although the management of these ecosystems is dependent on this knowledge (NRC, 2004a). Most instream flow programs specify a single, minimum value of stream flow that is required to (1) meet a legal standard or (2) sustain an endangered species or some other flow-dependent resource(s). However, current trends in instream flow programs are moving away from these single values and towards comprehensive river science. For example, instream flow hydrology and hydraulics now include the hydrologic regime with seasonal and inter-annual variation and not only a minimum flow value; biological aspects account for aquatic and riparian ecosystems and not just a single-species target species. In-channel and out-of-channel riverine physical processes are also considered, such as sediment dynamics and geomorphic processes, and water quality considerations include temperature, dissolved oxygen, nutrient loading, and toxics. In addressing stream flows across this broad spectrum of ecosystem conditions and processes, scientists now consider a fuller range of stream flow conditions beyond minimum instream flow needs.

This report recommends instream flow programs be designed to incorporate several key characteristics. First, instream flow programs need well-defined and measurable goals to frame instream flow studies and evaluate program progress. Clear goals are needed to increase efficiency and applicability of time- and resource-intensive technical evaluations. Stakeholder input in determining instream flow goals is important because there are usually many competing demands for water and competing opinions on how to allocate that water. Public support will be easier to garner when goals are easily measured and communicated.

Second, state-of-the-science programs use natural flow characteristics as a reference for determining flow needs. Natural river systems have variable flows (also called flow regimes) within a year and among multiple years. For example, in most Texas rivers, the lowest natural flows occur during warm, growing seasons of the summer and fall. During this same period there might also be some temporary high-flow peaks driven by storms, especially in those areas of the state subject to tropical storms. This natural variability is important to sustain aquatic and riparian biota and riverine processes.

Third, river science is not just for hydrologists anymore. Riverine science is now an inter- and multi-disciplinary science that includes biological, hydrological, geomorphic, and water quality aspects. Accordingly, successful instream flow programs will employ an interdisciplinary team of scientists to address the different elements of a river system. This team will include specialists in hydrology, biology, water quality, and physical processes who focus on whole functioning ecosystems and flow regimes.

Finally, a successful program will practice adaptive management in implementing instream flow recommendations over the long-term of the program. The processes of conducting instream flow studies will become better understood in Texas over the years it takes to complete the priority river basin studies and implement the flow recommendation(s). Some aspects of the current Texas programmatic approach may need to be modified as the results from the first studies are evaluated. Adaptive management is defined in the TOD as an "approach for recommending adjustments to operational plans in the event that objectives are not being achieved." Use of adaptive management will allow the agencies and other interested parties to test and revise the way that the instream flow program is implemented by assessing the ecological responses to new flow regimes. The adaptive management approach entails a long-term commitment to monitoring and anticipates corrections and revision over time.

EVALUATION OF THE PROGRAMMATIC WORK PLAN

The PWP makes clear that instream flow components—hydrology and hydraulics, biology, geomorphology, and water quality—form the core study elements needed to gain a minimal understanding of any river ecosystem. In crafting the PWP, the Texas agencies embraced an interdisciplinary approach that captures important aspects of instream flow studies consistent with the state-of-the-science. For example, the PWP explicitly includes a range of technical components and a multiple-step process. It also calls for monitoring to assure that the implemented flow regime meets study objectives and provides a basis for adaptive management.

Despite these strengths, the proposed instream flow program could be strengthened with revisions to the PWP. The PWP should be revised to: (1) define sound ecological environment, (2) assure statewide comparability with studies tailored to local conditions, (3) establish clearer goals, (4) embrace a two-step instream flow process, (5) modify the proposed flow chart, and (6) explain how indicators will be selected and used for specific river basins and statewide.

Sound Ecological Environment

The Texas instream flow program is predicated on legislative language in Texas Senate Bill 2 (2001) that directs the three Texas state agencies to "... conduct studies and analyses to determine appropriate...flow conditions [that]...support a sound ecological environment." A "sound ecological environment" is not defined in the legislation or the PWP. The meaning of a sound ecological environment ultimately will be reflected in all subsequent objectives, data collection, and analytical methods of the instream flow program. A clear definition of "sound ecological environment" will provide structure to the state's instream flow program and give context to the individual instream flow studies. **A clear definition of the phrase "sound ecological environment" needs to be provided to supply context for instream flows in Texas**.

State-wide Consistency and River Basin Specificity

Developing an instream flow program across a large and diverse state presents a special challenge. In Texas, the instream flow program is administered and overseen at the state-level, but instream flow studies are tailored for specific river basins. Therefore, the program must simultaneously establish methods specific enough to guide repeatable, technical evaluations at the subbasin scale and guidelines broad enough to apply to all rivers systems in Texas.

Consistency among individual studies at a high level will allow the state agencies to manage the instream flow program as a single program, not as a collection of basin-level instream flow studies. Basin-scale specific conditions can be accommodated in the individual studies that select methodologies and tools from state-sanctioned processes. This way, all methodologies used in the technical evaluations, regardless of subbasin characteristics, are approved at the state level so that results can be compared across subbasins, as applicable. Indeed, a statewide and state-sanctioned process for conducting individual studies would help ensure consistent method applications and consistent interpretation of instream flow recommendations. As written, the PWP provides a very limited structure to ensure consistent or comparable instream flow studies across the priority study sites. **The PWP should present a state-wide context for individual subbasin studies with two levels of oversight: one at the state level for management and program consistency and one at the subbasin level for goals and approaches that are tailored to the specific needs of the study basin.**

Goals

For both the state- and the basin-scales, the PWP needs more attention to the process of setting goals and the means to measure progress towards achieving those goals. Once "sound ecological environment" is clearly defined, goals can be established that will help riverine environments meet the criterion of "sound." State-level goals should define the objectives for the state's instream flow program and should encompass the broad-level milestones expressed in the legislative language of Texas Senate Bill 2. These programmatic goals should establish some of the parameters for the basin-level goals that will necessarily be more technical in nature. The PWP outlines one general goal of an instream flow study to "determine an appropriate flow regime…that conserves fish and wildlife resources while providing sustained benefits for other human uses of water resources." This goal does not give enough detail to guide consistent basin-level studies across the state and may actually generate conflict because conserving fish and wildlife and providing for human use may be mutually exclusive. Basin-level goals should guide the technical evaluations and be consistent with the state-level program goals. Means to set basin-level goals are not mentioned in the PWP or TOD. Program implementation and conduct of individual studies will be enhanced to the extent that clear, specific goals at the state- and basin-levels are consistent with a "sound ecological environment" and communicated with resource agencies, managers, scientists, and stake-holders. The PWP should present clear and specific goals for the state-wide instream flow program and recognize the need to develop individual sub-basin goals that nest within the state-wide instream flow programmatic goal(s).

Two-Phase Instream Flow Process

Setting goals and measuring success toward those goals are important steps in a larger, two-phase process for establishing instream flow recommendations. The first phase is the study design that includes a review of all relevant existing information and the conduct of reconnaissance studies, if necessary, prior to undertaking detailed (and potentially resource-intensive) evaluations. These initial assessments should describe the major processes and dynamics of the river's physical and ecological environment, identify specific questions to be addressed in the detailed technical evaluations, and inform the selection of methods to be used in the detailed technical evaluations. **The PWP and the TOD should describe how existing information and reconnaissance studies will be used to guide the detailed technical evaluations of hydrology and hydraulics, physical processes, biology, and water quality.**

In the second phase, detailed technical evaluations address the questions from the initial technical evaluations within one or more technical areas. Results from the initial and detailed technical evaluations should be (1) used within the river basin to derive proposed instream flow recommendations; (2) communicated to the state-level; and (3) integrated at the state-level such that statewide approaches for initial and detailed technical evaluations emerge.

Revised Flow Chart

A proposed flowchart (Figure ES-1) is a modified version of the PWP flowchart. The proposed flowchart emphasizes certain important steps in

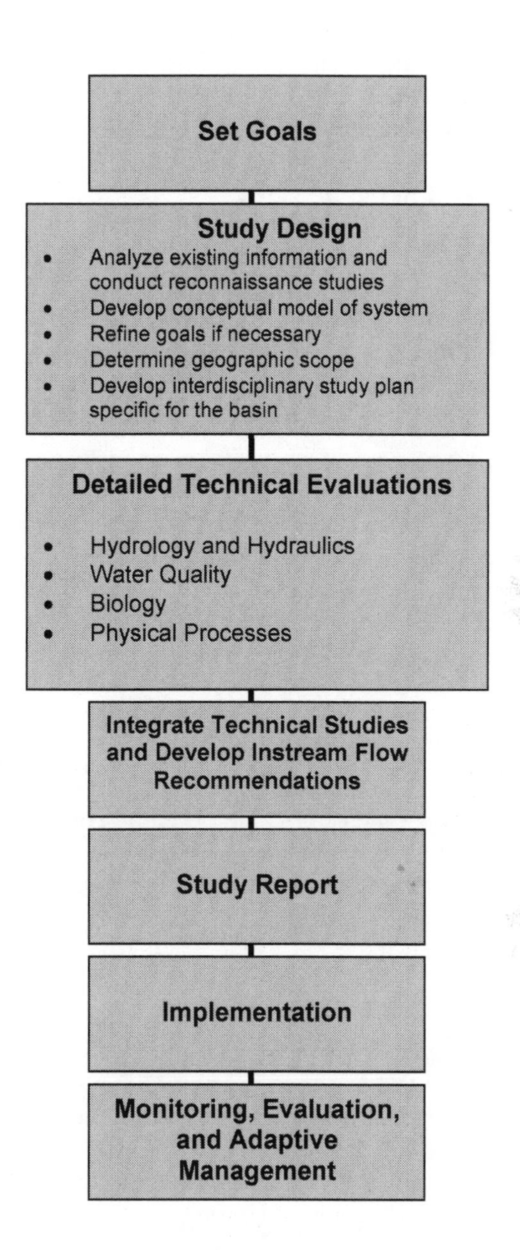

FIGURE ES-1 Recommended flowchart for instream flow studies.

conducting an instream flow study. The current PWP presumes goals and does not clearly articulate connections between existing information andreconnaissance studies and the detailed technical evaluations. **The PWP flowchart for instream flow studies should be revised to include several important steps in planning and conducting an instream flow study as suggested in Figure ES-1.**

Indicators

Indicators can measure progress towards achieving goals. Indicators related to flow characteristics could be used at the state-level in priority sites and in non-priority sites to identify and prioritize new studies. Once established, such indicators could be used to make quantitative comparisons among rivers segments. For example, the Lower Guadalupe River is considered more pristine than the lower San Antonio River, but this distinction has not been quantified. State-wide indicators, modified appropriately for regional differences, could also help track changes in the ecological conditions of Texas rivers over time in response to regulatory programs, such as the reduction in wastewater discharges from treatment plants or from management practices to address nonpoint sources.

At the basin-level, indicators are important connectors between basin goals and the instream flow recommendation. For example, if the basin goal is to increase the abundance of cottonwood trees (*Populus* spp.) in a riparian forest, then an indicator could be stem density of cottonwoods per unit area, and the flow recommendation would stipulate overbank flows at a certain level or frequency. In this case, the indicator is measurable and related to the flow recommendation, and adjustments could be made to the flow recommendation if the goal of increasing cottonwood abundance is not being achieved.

Developing accurate, reliable ecological indicators for the entire state will take several years. A workable and realistic set of indicators is likely to emerge only after several or all of the six priority instream flow studies have been completed. During the years required to conduct the priority studies, adaptive management methods should be employed to continually fine-tune ecological indicators through additions, deletions, and other changes. The PWP mentions the importance of monitoring and validation, but makes little reference as to how monitoring and validation would be conducted.

Texas has an example of successful indicators in its existing water quality monitoring programs. Bacterial and dissolved oxygen content in water are used as indicators that quantitatively support Texas' assessment and regulation of water quantity and quality. Like these indicators for water quality, a set of indicators is needed for the instream flow program and basin-scale studies. These indicators can be used in adaptive management, monitoring and validation activities to measure progress towards achieving a sound ecological environment in Texas rivers. **A suite of measurable, ecological indicators should be established for the state-wide program and each basin-specific study; the indicators should be responsive to instream flows.**

EVALUATION OF THE TECHNICAL OVERVIEW DOCUMENT

The TOD discusses sampling methodologies and modeling approaches proposed to conduct instream flow studies. Accompanying documents provide further detail on current Texas studies, processes to be considered, background information, and associated water-related programs, including information on the state total maximum daily load (TMDL) program. This study finds that the TOD appropriately identifies the relevant technical aspects of a comprehensive instream flow program (i.e., biology, hydrology and hydraulics, physical processes, and water quality) and mentions an approach to bring together these disparate elements (integration).

One strength of the TOD is its recognition of the importance of monitoring and validation, and the need for long-term, adaptive management. Adaptive management will be an important characteristic of an effective instream flow program, and the use of measurable indicators to monitor progress towards a sound ecological environment in Texas river basins is encouraged.

However, the TOD makes little distinction among individual basins and presents its methods as though each method is equally applicable across highly variable river basins. Furthermore, the TOD technical sections vary widely in quality and level of detail. Some sections present very detailed methods (e.g., the sections on hydrology and hydraulics and biological sampling), but other sections have little or no detail on the methods to be used, and others have significant omissions of important information. Rarely are methodologies presented in the TOD such that an instream flow recommendation could easily emerge. None of the technical sections refer to basin goals or a sound ecological environment.

The TOD discusses technical methodologies by discipline (i.e., biology, hydrology, etc.) and as separate studies, but does not describe how studies in different disciplines relate to each other or relate to an instream flow recommendation. This report suggests ways to connect various biological, hydrologic, and physical processes with water quality technical studies to create an instream flow recommendation. The various technical assessments are recommended to be framed in terms of flow regime components: subsistence flows, base flows, high flow pulses, and overbank flows (see Table 3-2). With the technical evaluations presented in terms of flow, connections will be strengthened among the evaluations and between the evaluations and the flow recommendation.

The TOD needs significant revision to reflect (1) site-specificity at the (sub) basin-scale; (2) goals for the individual studies that relate to the definition of a sound ecological environment; and (3) linkages among individual studies of biology, hydrology and hydraulics, physical processes, and water quality.

The **hydrologic and hydraulic** section of the TOD reflects a significant understanding of hydrology, and hydrologic measurement and analyses commonly required for performing instream flow studies. To be efficient in hydrologic and hydraulic analyses and to avoid performing analyses that are either not necessary or are more detailed than is needed for making instream flow recommendations, hydrologic and hydraulic approaches should be closely aligned with the other technical evaluations and with the goals for the specific river basin.

The strengths of the **biology** section include a strong general discussion of the important issues of habitat scale, ecological processes, and species life histories. The biology section of the TOD provides highly detailed accounts of how to conduct some sampling or modeling methods, but gives scant attention to how modeled and empirical data are communicated, related to program goals, or integrated with other aspects of an instream flow study to derive a flow recommendation.

The TOD captures the importance of **physical processes** in forming the channel and floodplain and in providing habitat for aquatic organisms, but the physical process section needs augmentation to be consistent with the content depth and quality in the hydrology and hydraulics and biology sections. It also needs to discuss hydrologic regimes common in Texas rivers, GIS applications, sediment budget methods, and impacts of land use, population, and climate change in the watershed as relevant aspects of riverine physical processes.

The TOD ably describes the **water quality** programs in Texas. Instream flow considerations are not the focus of the state's water quality programs. Therefore, the instream flow program's elements that describe water quality must be aligned with the existing water quality programs, so as to avoid conflicting requirements for maintaining sound ecological environments in Texas rivers. A significant limitation of the water quality section of the TOD is that it does not outline how the water quality component of an instream flow assessment should be conducted or how instream flow and water quality considerations can be integrated with each other.

Scaling Issues

The physical, chemical, and biological processes of a stream ecosystem operate at different spatial scales and are expressed differently over different time periods. In instream flow work these different spatial and temporal scales must be reconciled so that integrated, individual studies can be conducted to derive a flow recommendation. At present, the TOD does not specify what length of a river must be studied, how study reaches are selected, or how data from study areas will be extrapolated to unstudied areas. These shortcomings of the TOD are non-trivial and not easy to address. Scaling issues remain a major research focus for instream flow science, and effective methods for reconciling different scales are not well documented. Despite the difficulty in doing so, the various components of a study need to be compatible in terms of spatial scale. **Overall, the biological, physical processes, water quality, and hydrology and hydraulics instream flow studies should be designed at commensurate spatial and temporal scales to improve the ability to integrate findings from the various technical evaluations into a single flow recommendation.**

Integration

Integration is the process of combining the different technical components of instream flow studies into a flow recommendation. Integration is an important, complicated step in instream flow science, and while integration methods are being generated empirically, they are not well documented in the literature. The TOD presents a different way of doing integration at the end of the study process, where the results from the detailed technical evaluations are used to derive a flow recommendation. The TOD presents an integration framework (Figure 5-1) diagram to illustrate integration, but this diagram is complicated and not thoroughly explained. Thus, how results of the individual studies are to be combined into a recommendation is not clear in the TOD. **The TOD integration framework needs to be revised to include sequential steps and clearer direction of how to derive flow recommendations from the results of the technical evaluations.**

PROGRAMMATIC ISSUES

Linkage to Other Texas Programs

Several water-related programs already exist at the state-level, including those associated with water quality, stream flow, bays and estuaries, and water permitting. The instream flow program can build upon or augment existing, related water resources programs in Texas, and potentially share data, methods, and procedures with those programs. For example, Texas collects state-wide data on temperature, dissolved oxygen, and other chemical constituents, as well as biology, as part of its water-quality program. In this program, four levels of aquatic life use are defined (exceptional, high, intermediate, and limited). The Texas Administrative Code establishes water quality aquatic life use goals for all 225 classified stream segments. At a minimum, the existing aquatic life use goals could be considered in implementing instream flow recommendations to avoid conflict or establish support between the instream flow and water quality programs.

Integrating the instream flow program with existing water quality and quantity programs will provide clear and consistent direction for both decision makers and stakeholders. Streamlining related programs will also reduce the potential for inconsistent recommendations among the programs, reduce costs, and eliminate redundant analyses. **The instream flow program should be integrated with the water quality, water permitting, and other water-related programs in Texas**.

Peer Review

Maintaining scientific excellence in the Texas instream flow program could be facilitated with access to and open communication with technical experts from instream flow-related disciplines. An important role for reviewers is to evaluate the results and methods of the individual technical studies, as well as the progress of the overall instream flow program development. Results from these reviews should be communicated to the scientists involved in the Texas studies, the instream flow scientific community at large, and stakeholders. Review by an independent group of scientists will help track the progress and efficacy of the instream flow program over time, just as the initial peer review was designed to provide, "the highest level of confidence... that the framework [for]... these studies... is scientifically sound" (TPWD, TCEQ, and TWDB, 2002). In order to fulfill this comprehensive program objective that involves scientists from a variety of disciplines, state agencies, and other stakeholders, **the creation of an independent, interdisciplinary, periodic peer review process for the instream flow program is recommended.**

Implementation Issues

This report focuses on the scientific and technical aspects of the Texas instream flow program as presented in the PWP and TOD. Nevertheless, several practical implementation issues arose during the course of this study. The act of implementing an instream flow program or study requires deft balance among disparate and competing uses for river water. Large-scale, state-wide instream flow programs, like the one in Texas, are often implemented over a numbers of years. Over the life of the Texas instream flow program, and through adaptive management, many changes may be made to instream flow methodologies, implementation, or goals of the program. The Texas instream flow program has identified six priority river basins to initiate the instream flow program. These priority basins represent a small subset of the total number of rivers and streams in the state, and the state may wish to expand the instream flow program to other rivers as it develops instream flow experience. Preserving the status quo, especially on important rivers, may be important at least until the initial period is over and focus can be turned to non-priority river systems' instream flow requirements. Ideally, a priority-setting methodology would help water managers determine the order in which additional rivers will be evaluated for instream flow recommendations and weigh a range of alternatives to maximize the state's future opportunities to protect adequate instream flows.

CONCLUSIONS AND MAJOR RECOMMENDATIONS

Developing instream flow recommendations for rivers is one of the most difficult and important challenges in the applied ecological and physical sciences today. The Texas agencies are commended for proposing a prospective, comprehensive instream flow program. Implementation of a statewide instream flow program will involve many agencies, significant resources, and time; nevertheless, the program will provide enormous benefits to the state over the next several decades and beyond.

The Texas instream flow program will need to be flexible to meet the unique challenges and opportunities presented by the state's rich mixture of river ecosystems, culture, water law, and water development. Clear and specific programmatic and scientific instream flow goals need to be set at both the state and river basin levels, and methods used in setting instream flow recommendations need to be consistent for the several river systems that will be studied across the state. The Texas instream flow framework should elicit comparable results at the basin level in order to realize state-wide consistency, maintain continuity over the long term through proper delegation and delineation of responsibilities among the various involved agencies, and incorporate scientific findings as well as social and economic concerns by involving stakeholders during key phases of the design and implementation process.

Major Recommendations

1) The PWP should present a state-wide context for individual sub-basin studies. This can be accomplished with two levels of oversight: one at the state level for management and program consistency and the second one at the subbasin level for goals and approaches tailored to the specific needs of the study basin.

2) A clear definition of the phrase "sound ecological environment" needs to be provided to supply context for instream flows in Texas.

3) The PWP should present clear and specific goals for the state-wide instream flow program and recognize the need to develop individual sub-basin goals that nest within the state-wide instream flow programmatic goal(s).

4) The PWP and the TOD should describe how existing information and reconnaissance studies will be used to guide the detailed technical evaluations of hydrology, physical processes, biology, and water quality.

5) The PWP flowchart for instream flow studies should be revised to include several important steps in planning and conducting an instream flow study as suggested in Figure ES-1.

6) A suite of measurable, ecological indicators should be established for the state-wide program and each basin-specific study; the indicators should be responsive to instream flows. These indicators can be used in adaptive management, monitoring and validation activities to measure progress towards achieving a sound ecological environment in Texas rivers.

7) The Technical Overview Document should be revised to provide for consistent spatial scale and level of detail for the hydrology, biology, physical processes, and water quality technical evaluations.

8) Clearer direction should be provided for the process by which the individual technical evaluations will be integrated into instream flow recommendations.

9) The instream flow program should be integrated with the water quality, water permitting, and other water-related programs in Texas.

10) The creation of an independent, interdisciplinary, periodic peer review process for the instream flow program is recommended.

1

Introduction

TEXAS WATER RESOURCES

Water resources in Texas have been important in the state's history, settlement and current economic development. Most of the state's boundaries are defined by rivers: the Sabine River on the east, the Red River to the north, and the Rio Grande along the southwestern border with Mexico (see Figure 1-1). Within Texas, several large rivers traverse the state, generally flowing from the northwest to the southeast and emptying into the Gulf of Mexico. Texas rivers such as the Brazos, Pecos, and Trinity have served as important transportation arteries and are part of historical lore in Texas and the American West. The state also contains numerous other streams that serve as sources of water for urban populations and provide important water supplies for riverine ecosystems. In some parts of the state, especially its arid western portions, groundwater supplies have long served as important sources of water for livestock and, more recently, as sources of water for irrigated agriculture.

Like many parts of the southern and western United States, Texas experienced marked population growth during the 1980s and 1990s. The state registered a sizable 22.8 percent growth from 1990-2000, and its 2003 total population was estimated at over 22 million, second only to California's total population[1]. Such growth is projected to continue, as estimations suggest that by the year 2050 as many as 900 Texas cities will need to reduce water use or develop new supplies to meet demands during drought periods (TWDB, 2002a). Population growth and associated increasing urban demands occur simultaneously with other Texas water supplies and demands: limits on the abilities to develop new supplies or re-allocate water among existing users; legal obligations to provide flows to sustain species and habitat; and greater demands for flows to support recreational, aesthetic, and related preferences. This dynamic setting is straining the ability of Texas rivers and streams to meet these sometimes competing demands. The three state agencies responsible for water resources in Texas are the Texas Water Development Board (TWDB), the Texas Parks and Wildlife

[1] Data from *http://quickfacts.census.gov/qfd/states/48000.html.*

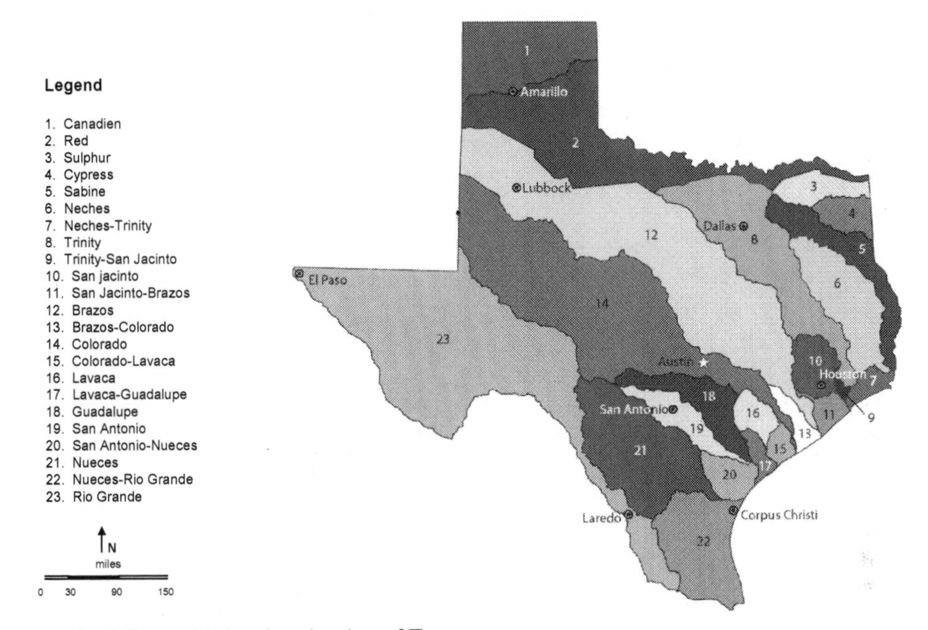

FIGURE 1-1 Major river basins of Texas.
SOURCE: Adapted from Hayes, 2002.

Department (TPWD), and the Texas Commission on Environmental Quality (TCEQ). These three agencies are also challenged to define the state's streamflow and related water management policies.

During the 1980s and 1990s, these three Texas state agencies began to develop programs designed to provide specific flow rates or "instream flows," in Texas streams and rivers in order to balance competing needs for limited flows. In Texas, instream flow describes "a flow regime adequate to maintain an ecologically sound environment in streams and rivers including riparian and floodplain features and necessary for maintaining the diversity and productivity of ecologically characteristic fish and wildlife and the living resources on which they depend" or flows needed to "support economically and aesthetically important activities … [including] navigation" (TPWD, TCEQ, and TWDB, 2002).

TEXAS INSTREAM FLOWS PROGRAM

The Texas Instream Flows Program has its roots in two State Senate Bills. Senate Bill 1 (1997), commonly referred to as the "Water Bill," established the state water planning process with provisions for environmental values to be considered in water development and transferal activities. Senate Bill 2 (2001) takes the state water planning process further and initiated the instream flow program. Specifically, this Bill directs the TWDB, the TPWD, and the TCEQ to "jointly establish and continuously maintain an instream flow data collection and evaluation program," and to conduct studies that determine flow conditions in the state's rivers and streams necessary to support a "sound ecological environment." Senate Bill 2 stipulates that priority studies are to be completed no later than December 31, 2010.

In response to Senate Bill 2, the three Texas state agencies designed the state instream flow program and present it in two documents, the Programmatic Work Plan (PWP; TPWD, TCEQ, and TWDB, 2002) and the Technical Overview Document (TOD; TPWD, TCEQ, and TWDB, 2003). The PWP outlines the scope, timeframe, and methods that the agencies are proposing to plan, design, and implement priority flow studies. The PWP identifies the goals of an instream flow study to "determine an appropriate flow regime (quantity and timing of water in a stream or river) that conserves fish and wildlife resources while providing sustained benefits for other human uses of water resources." Eight components give structure to Texas instream flow studies: study design, hydrology and hydraulics, biology, physical processes, water quality evaluations, integration and interpretation, study report, and monitoring and evaluation activities. For every study, the three state agencies are proposing to divide and share responsibilities among the eight elements, depending on expertise. The TOD describes the technical aspects of instream flow studies, including sampling methods for individual technical evaluations.

The Texas instream flow program design has three phases. Phase one is the drafting of the PWP and the development of the TOD (completed December 2002). The second and third phases are peer-review activities. Phase two (this National Research Council (NRC) study) entails an objective, third-party review and evaluation of the scientific basis and soundness of the scientific and engineering methods proposed for use in Texas instream flow projects. Phase three is continued peer-review by Texas river authorities and stakeholders impacted by instream flow water management decisions.

THE NRC STUDY

In early 2003, the TWDB requested the NRC's Water Science and Technology Board to review the program and technical methods proposed for establishing Texas instream flow recommendations. Later that year, a committee of experts was appointed to evaluate the scientific methods, materials, and related technical aspects of the proposed PWP and TOD for developing instream flow studies in Texas. The committee conducted its deliberations and issued its report in accordance with the task statement contained in Box 1-1.

The committee met three times in Texas between autumn 2003 and spring 2004 in Austin, San Antonio, and San Marcos. A fourth and final meeting was convened in Washington, D.C. in May 2004. Portions of the first three meetings included sessions that were open to the public, and the committee heard from a wide range of experts and citizens with interests in the Texas instream flow program and in this study. People were also invited to submit written comments for the committee's consideration. Many individuals accepted this invitation, and these written comments were considered along with formal presentations.

The NRC study and this report are directly responsive to the needs and the request for assistance of the three Texas agencies, but this report may apply to instream flow issues beyond the borders of Texas. In Texas and other western states, demands of growth tax water supply and quality and fair water appropriation. Texas water issues are microcosmic of national water issues: uneven distribution across space and time, competing uses, increasing demands, and changes in social preferences. Therefore, Texas' approach to instream flows may serve as a guide for other jurisdictions wrangling with similar problems.

The study's statement of task (Box 1-1) defines the scientific and technical issues associated with instream flows in Texas that are considered in this report, and the report reflects that charge. This report provides rigorous evaluations of the PWP and TOD. During the course of committee deliberations on the scientific and technical dimensions of instream flows, the context in which such flows are implemented emerged as being very important. The report thus comments on the scientific aspects of instream flows and the decision-making context in which instream flow recommendations are implemented.

ORGANIZATION OF THIS REPORT

The report is presented in five additional chapters. Chapter 2 introduces the necessary context for instream flow studies in Texas, including a

BOX 1-1
Statement of Task for Texas Instream Flows

The committee will appraise the scientific and engineering methods used to help establish instream flow recommendations in Texas rivers, and focus on the soundness and adequacy of the Programmatic Work Plan for developing instream flow studies developed by the TWDB, TCEQ, and TPWD. Specifically, the NRC committee will:

1. Evaluate the key documents that explain these scientific and engineering methods and their applications in setting instream flow recommendations. These documents are a) the 2002 Programmatic Work Plan, and b) a supplementary technical volume that describes these methods in greater detail.
2. Review and provide advice on several scientific and technical matters relevant to instream flow studies and recommendations, including:
   a. appropriate spatial scales of analyses in hydrologic and related models;
   b. use of habitat-flow relations in setting instream flow requirements;
   c. use of landscape ecology metrics in setting instream flow requirements;
   d. range of biophysical model parameters employed in the Texas State TMDL program;
   e. applicability of water quality models used in the Texas State TMDL program to instream flow studies.
3. Evaluate findings and recommendations of Tasks 1 and 2 for consistency with the requirements of Texas law for the study of instream flows

description of Texas river environments across the large state, current efforts and agency programs that provide the programmatic context for the instream flow study program, and Texas water code and legislation that frame the instream flow program. A brief tutorial on instream flow science and concepts is presented in Chapter 3, including examples of instream flow studies that have been implemented. The instream flow tutorial briefly discusses the scientific bases for instream flows and the characteristics of the most effective studies.

Chapters 4 and 5 present the evaluations of the Texas PWP and accompanying TOD, respectively. The evaluation of the PWP focuses on general plans for the program as a whole as well as plans for individual river basin studies. The TOD is evaluated in its entirety and by discipline: hydrology and hydraulics, physical processes, biology, and water quality, and integration of separate discipline studies into an instream flow recommendation. These evaluations of the Texas instream flow documents constitute the bulk of the report, findings, and recommendations. The final chapter of the report focuses on implementation aspects of the instream flow program, challenges of implementing an instream flow program in Texas, and integration of the instream flow program with existing water-related state programs.

2

Scientific and Program Context for the Texas Instream Flow Program

The Texas instream flow program exists within scientific and program contexts. The scientific context includes the state's hydrologic, physical, and climatic settings; the program context includes water management statutes and Texas water programs. This chapter describes these contexts of the Texas instream flow program.

SCIENTIFIC CONTEXT

Texas climate and topography exhibit great variations across its vast 266,805 square miles. Topography across the state includes flat, level plains of the Texas Panhandle, basins and mountains in the Trans-Pecos region, and rolling hills in east Texas. Western areas of the state are dry and the coastal areas in the east are humid. Texas' wide span of hydrologic and physical riverine conditions impacts how instream flow science is conducted across the state.

Hydrologic Setting

Precipitation ranges from an average of 8 inches per year in far West Texas to as much as 60 inches per year in coastal east Texas (TPWD, TCEQ, and TWDB, 2003). Texas rivers reflect this precipitation variability: rivers in west Texas generally exhibit greater seasonality in flows and a higher frequency of flash floods, and rivers in east Texas generally carry higher flows with less seasonal variation. Many of the state's streams and rivers flow from the north and west toward the south and east (see Figure 1-1). Texas, more than other states in the United States, has a hydrological regime with a high flash-flood potential (Beard, 1975). This potential varies across the state from west to east, like the river drainage basins themselves (see Figure 1-1), and it is an important consideration in the hydrology, hydraulics, and aquatic ecosystems in Texas rivers.

Differences in hydrologic regime across the state have important implications for instream flow science. For example, in the flood-dominated river basins of central and west Texas, geomorphic dynamics do not conform to the classic "equilibrium" concepts of geomorphology, even when land use change, channelization, or dam construction have not occurred in the watershed (Baker, 1977). Therefore, geomorphological dynamics of rivers in Texas follow a west-east spatial gradient across the state from disequilibrium behavior in the west to equilibrium-like behavior in the east. This strong hydro-geographic gradient is reflected in the physical structure of aquatic and riparian habitat and other ecological processes and patterns.

Physical Setting for Instream Flows

The physical setting of Texas rivers can be described in many ways for the wide variety of conditions across the large state. For descriptive, instream flow purposes, the state of Texas and its river systems are coarsely categorized into five generalized districts: East, North-Central, South-Central, Lower Rio Grande basin, and West. These districts are described briefly below in terms of geology, climate, hydrologic regime, and biota.

East Texas

East Texas rivers (Lower Red, Lower Trinity, Lower Brazos, Navasota, Sabine, Neches) drain the portion of Texas with average rainfall between 30 and 50 inches a year. The region is dominated by flat landscapes and either clay-rich or sandy soils (the latter associated with the Sabine and Neches watersheds). Rivers of this region historically experienced periodic flood pulses that connected river channels to floodplains. Watersheds of the region are dominated by agriculture. The 1950s were a period of dam construction across this region, and today most major rivers have been impounded for flood control purposes. Water-based recreation is popular in this region, especially fishing in some of the state's largest and most productive reservoirs. The region contains several imperiled aquatic species, including paddlefish (*Polyodon spathula*) and sharpnose shiner (*Notropis oxyrhynchus*). Fish communities in east Texas (in basins like the Brazos River) are dominated by species adapted to high variations in flow, high turbidity (especially in the Trinity, Brazos, and Red drainages), and harsh environmental conditions. Channel substrates are mostly soft, shifting sediments (sand, mud, and silt). The dominant physical structure within stream channels is woody debris.

North-Central Texas

North-central Texas rivers (Canadian, Upper Brazos, Upper Colorado, Upper Red, Upper Trinity) drain watersheds with clay-rich soils and heavy agriculture use. This region is drier than east Texas, with rainfall averaging between 15 and 28 inches a year and occasional severe droughts. These rivers have flow characteristics similar to those of East Texas rivers, but they are smaller and tend to experience more frequent drought conditions. The region is dominated by fish species that are resistant to alternating drought and flood conditions. Like east Texas, water-based recreation is also quite popular. This region includes several of the state's large metropolitan areas (Dallas/Forth Worth, Amarillo, Lubbock, Waco).

South-Central Texas

This region is better known as "The Hill Country" of Texas. Rivers that drain this region include the Blanco, Comal, Frio, Guadalupe, Lower Colorado, Nueces, Sabinal, San Antonio, and San Marcos rivers. The landscape in this region is rocky in many areas, and the dominant land use in this region is livestock grazing. The region, which includes the Edwards Plateau, has a relatively wide range of average annual precipitation. Parts of this region are relatively dry and experience periodic drought, with average annual rainfall of around 10 inches, other parts receive up to about 40 inches a year. Rivers in the region receive significant subsurface flow and tend to flow clear and cool most of the time, but experience relatively infrequent flash floods during spates. The region harbors several threatened and endangered fishes including the fountain darter (*Etheostoma fonticola*), Clear Creek gambusia (*Gambusia heterochir*), and cave catfishes (*Satan eurystomus, Trogloglanis pattersoni*). Two of the state's fastest growing metropolitan areas, Austin and San Antonio, are located along this region's eastern border. The rapid population growth of these two urban areas has placed sharp demands on the region's limited water resources. Hill Country rivers and streams are used for a variety of recreational purposes, including swimming, rafting, canoeing, and fishing.

Lower Rio Grande Basin (Lower Rio Grande, Devils)

The largest rivers in south Texas are the Lower Rio Grande River and its tributaries, including Devils River, Las Moras Creek, and San Felipe Creek. Annual average rainfall in this region varies from 11 to 26 inches. The region's prevalent land uses are irrigated row cropping in the Lower Rio Grande Valley, and livestock grazing across the region. The region's major cities are Brownsville, McAllen, and Del Rio. Population growth in this region also is exerting increasing pressure on limited water resources. Over the past several decades, instream flow in the Lower Rio Grande has been progressively reduced by upstream water diversion, withdrawal, and evaporation from reservoirs. Today, the Lower Rio Grande channel is periodically reduced to a series of isolated pools, and the river fails to reach the Gulf of Mexico for extended periods. The Rio Grande is an extreme example of how aquatic biota evolutionarily adapted to pre-Columbian stream flows have been stressed to the point of being detrimental to their survival by changes and disruptions to natural flows. Threatened and endangered aquatic species in this region include the Devils River minnow (*Dionda diaboli*), prosperine shiner (*Cyprinella proserpina*), and Rio Grande darter (*Etheostoma grahami*). Water-based recreation use is increasing on those rivers that have more reliable year-round flows.

West Texas

West Texas is the driest region of the state. Some areas of west Texas receive annual average rainfall of roughly 16 inches, but that figure falls to less than 8 inches in far west Texas. The region's aridity has resulted in strong pressures on its surface and groundwater resources. The principal land use is livestock grazing, especially for sheep and goats. The principal rivers in this region are the middle Rio Grande and the Pecos. The Pecos River is highly saline and has experienced golden algae blooms that kill fish and other aquatic animals (Rhodes and Hubbs, 1992). Endangered aquatic species are a common occurrence and include the Comanche Springs pupfish (*Cyprinodon elegans*), Leon Springs pupfish (*C. bovinus*), Pecos pupfish (C. pecosensis), and Pecos gambusia (*Gambusia nobilis*).

Climate

Climate, which controls hydrology and affects all other aspects of a river system, is a critical element in water planning in Texas (TWDB, 2002a). The U.S. National Assessment of climatic change impacts on the U.S. (U.S. National Assessment Synthesis Team, 2001) reports that the southern Great Plains is likely to experience increases in temperature, frequency of heat-stress events, and precipitation changes with a shift toward more intense rainfall events, and frequency and severity of droughts. The result is "expected to exacerbate the current competition for water among the agricultural sector, urban and industrial users, recreational users, and natural ecosystems, as well as within each user community" (U.S. National Assessment Synthesis Team, 2001).

PROGRAM CONTEXT

In 1840, Texas adopted the riparian doctrine that entitles a landowner to a reasonable use of river water that abuts his or her land. Thus, for much of the last half of the nineteenth century, land ownership determined the right to use river water in Texas. A series of state laws adopted at the end of the nineteenth century declared that then unappropriated surface waters were to be the property of the public and that future rights to use water would be acquired under a prior appropriations system. While Texas continued to honor pre-existing riparian rights, this new legal structure set a very different course for Texas.

Under the prior appropriation doctrine, the most senior water right holder is entitled to have his or her water right fully satisfied in times of shortage, before the next most senior water right holder is allowed to divert or store water. Thus, the prior appropriation system often is described as "first in time, first in right." The prior appropriation system typically limits appropriators to the diversion or storage of water for "beneficial uses," a term which has evolved in Texas and the western states as societal expectations have changed (for example, many western states now treat non-consumptive instream uses to be beneficial). Unlike the law of riparian rights, the law of prior appropriation also allows a water rights holder to divert water from a stream and to transport the water to be used at some distance from the stream, potentially even in a different basin. Finally, in Texas and other western states, a water right can be lost through non-use over a period of years.

Texas embraced this dual system of water law—riparian rights co-existing alongside appropriative rights—until it was merged with prior appropriation with the Water Rights Adjudication Act. The 1967 legislation merged the riparian system with the prior appropriation system that controls water allocation in most of the West by mandating a comprehensive adjudication of all water rights on individual river and stream segments. All pre-existing water rights that could be proved up were granted certificates of adjudication while future applicants were required to secure a permit, known as a water right, from the state agency, now the Texas Commission on Environmental Quality (TCEQ), charged with administration of the permit system. Since the inception of the current permit system, the TCEQ has also had the authority to grant, deny, and condition water rights to best serve the public interest.

Water Management Statutes

Several statutes adopted in 1977 expanded the TCEQ's ability to protect environmental values, including instream values, as part of its mandate to manage and allocate water resources in Texas. A state statute gave the commission the authority to maintain a proper ecological environment in the state's bays and estuaries for permits issued within 200 miles of the coast and the ability, when practicable, to include conditions that are necessary to maintain beneficial inflows to bays and estuaries. Other provisions allowed the commission, when it weighs applications for new and amended permits, to consider a diversion's effects on existing instream uses, water quality, and fish and wildlife habitat.

Prompted by a serious drought as well as dramatic projections of population growth, in 1997 the Texas legislature adopted a sweeping reorganization of its water resource management regime. The legislation, known as Senate Bill 1, mandated that the Texas Water Development Board (TWDB), TCEQ, and the Texas Parks and Wildlife Department (TPWD) work together to produce a state water plan that was, in turn, to be based upon water plans developed by sixteen regional planning organizations along with the TWDB's analysis and policy recommendations. The first state water plan, based largely upon the regional plans, was completed in 2002 and the state is now engaged in a second round of planning to refine that plan. Senate Bill 1 also enacted new provisions regarding a wide range of other difficult issues, including groundwater management, inter-basin transfers, reuse, water marketing, and cancellation of water rights for non-use. Finally, Senate Bill 1 also was described by then-Lieutenant Governor Bullock as having recognized that "water must be available to satisfy environmental needs for Texas's fish and wildlife habitat, instream flows, bays, and estuaries. This legislation adds numerous new provisions that require environmental water needs to be considered whenever a water supply is developed, transferred, reused, or marketed."

Four years later, in 2001, the Texas state legislature adopted another piece of legislation, Senate Bill 2, dealing extensively with water law. This legislation established a Water Advisory Council, and made a number of technical changes to the state's water code. Of particular importance for this report, the legislature also directed the three resource agencies—TCEQ, TWDB, and TPWD to:

- Jointly establish and maintain an instream flow data collection and evaluation system;
- Conduct studies to determine appropriate methodologies for determining flow conditions in the state's rivers ands streams necessary to support a sound ecological environment;
- Complete priority studies by December 31, 2010; and
- Direct the Commission to consider the results of the studies in its review of management plans, water rights, and interbasin transfers

Finally, it is important to note that much of the State of Texas's groundwater resources are managed under an allocation regime that is largely separate and distinct from the prior appropriations system used for surface waters. Texas law presumes that all underground water sources are "percolating waters," which are subject to the English common law doctrine known as the Rule of Capture. Within this legal framework, the land owner is generally permitted to withdraw water (to reduce water to his or her possession) from these underground sources even if the withdrawals deleteriously affect the quantity of water found beneath an adjacent surface property or diminishes the flow of surface streams. However, that rule is subject to limitation if the withdrawal causes malicious injury to another landowner, or causes willful waste.

Moreover, the Texas state legislature has established some general as well as some site-specific statutory constraints on the operation of the Rule of Capture. At the site-specific level, Texas has established a coastal subsidence district to limit pumping from the Gulf Coast aquifer in the Galveston region, and an aquifer authority which is charged with managing groundwater withdrawals from the Edwards aquifer to protect endangered species and maintain flow levels in the Guadalupe River. More generally, the Texas state legislature in Senate Bill 1 reaffirmed that local groundwater conservation districts are principally responsible for managing groundwater resources. It also provided the districts with more statutory authority than had previously been available. In addition, the Texas state legislature required these entities to develop groundwater management plans and the TWDB to certify the plans once they are complete. There are now almost eighty groundwater districts in Texas, most of which were created along county rather than aquifer boundaries.

Texas Water Programs

There are three Texas programs that deal with water availability and water quality that are directly relevant to conducting instream flow studies and implementing the results of those studies. Water availability is important because the amount of water in streams that remains available for allocation will have important effects on the ability of the Texas agencies to implement any flows recommended as a result of an instream flow study. The water quality programs are specifically mentioned in the Technical Overview Document (TOD) as producing information that should be considered in designing and conducting an instream flow study. These programs are briefly described below.

Water Availability

Permits are required to withdraw water from Texas streams and rivers. These permits are administered by the Texas Water Rights Permitting Program within the TCEQ. In response to Senate Bill 1, TCEQ made substantial improvements in the rigor of its evaluation of permits for surface water withdrawals by developing a water availability model (WAM) for each of the twenty-three river and coastal basins of Texas. That model is used both in the permitting process and in regional water planning. The input data set for the WAM identifies a set of control points where a control point is a diversion or storage location on a river, and includes physical data about each control point, such as the amount of permitted withdrawal, upstream drainage area, and the priority date. Also included in the input data set is a sequence of "naturalized flows" derived from United States Geological Survey (USGS) stations (the naturalized flow is a modeled flow which would have occurred if no diversions or upstream reservoirs existed). Typically, this flow sequence is defined for monthly flows and encompasses 40-50 years of historical data.

It appears that many permits have not yet been exercised, or have not yet been exercised in full. Water remains available for allocation in many streams and rivers across Texas. TCEQ defines the limits of water withdrawals from rivers using a specific percentage of naturalized flow as the minimum flow. Generally, water is less available in the upper parts of the basins than in the lower basins.[1]

Texas Pollutant Discharge Elimination System

The State of Texas implements the National Pollutant Discharge Elimination System (NPDES) program through the Texas Pollutant Discharge Elimination System (TPDES) and it applies to all point sources, including municipal effluent. In many cases, the agency must conduct a receiving water assessment before it can issue a TPDES permit, and Texas uses the QUAL-TX[2] model to provide analytical support for the TPDES program. This water assessment process using QUAL-TX creates conservative estimates of pollutant loads because QUAL-TX accounts directly for point-sources and indirectly for non-point sources. The biological studies needed to support the Texas instream flow program effectively require collection of much the same kind of data as is needed to complete a receiving water assessment of a wastewater discharge. However, the receiving water assessment is focused on the point of discharge and resulting effects downstream, while an instream flow study may need to characterize the entirety of a long river reach.

Texas Water Quality Inventory

The Texas Water Quality Inventory is prepared by TCEQ and submitted to the Environmental Protection Agency (EPA) biannually in even-numbered years in accordance with section 305(b) of the Clean Water Act. Water bodies that do not support their water quality standards and for which existing controls are not adequate are placed on the 303(d) list of impaired water bodies, and then come under the domain of the total maximum daily load (TMDL) program (discussed in detail in Chapter 5). For aquatic life use, the criteria include dissolved oxygen, physical habitat, toxic substances in water and sediment, and biological assessments, though ade-

---

[1] Data available on-line at *http://www.tnrcc.state.tx.us/permitting/waterperm/wrpa/wam.html.*

[2] QUAL-TX is a modification to the federal QUAL2E model that includes parameters specific to Texas rivers, such as a site specific equation for stream reaeration. The QUAL model was originally developed in Texas and later further developed and adopted for national use by the Environmental Protection Agency.

quate data to reach conclusions on these assessments are available for only a portion of all the water bodies in the state, especially for toxic substances and biological assessments. Aquatic life in Texas streams and rivers is to some degree impacted by toxic chemicals but the main water quality limitation on aquatic life is depressed dissolved oxygen.

3

An Introduction to Instream Flow Science and Programs

In the simplest terms, instream flow is the water flowing in a stream channel (IFC, 2002). This simple concept belies the difficulty of determining what that flow should be among competing uses for water, such as irrigation, public supply, recreation, hydropower, and aquatic habitat. The simple definition may not account for variations in flow levels across different seasons and wet, dry, and normal years. A challenge facing natural resource managers is to find a workable balance among these demands and use appropriate methods to quantify instream flow needs for each of these uses. Instream flow programs were created to meet this challenge.

An instream flow recommendation will give a numerical answer to the question, "How much water should be in the river?" Instream flow programs help water managers meet management goals of biology, municipal water supply, or water quality considerations. The Instream Flow Council (IFC) offers this definition for instream flows (IFC, 2002):

> The objective of an instream flow prescription should be to mimic the natural flow regime as closely as possible. Flow regimes must also address instream and out-of-stream needs and integrate biotic and abiotic processes. For these reasons, inter- and intra-annual instream flow prescriptions are needed to preserve the ecological health of a river.

Two primary literature sources describe instream flow science, *Instream Flows for Riverine Resource Stewardship* (IFC, 2002) and *Rivers for Life: Managing Water for People and Nature* (Postel and Richter, 2003). These books are based on instream flow research and studies conducted on many rivers in North America and the rest of the world and reference hundreds of citations. Information in these books, as well as other primary and secondary references, is used as a foundation for some of the conclusions and recommendations in this report.

This chapter offers a brief tutorial on the basic structure of instream flow science, studies, and programs. Trends and principles in the science are discussed, major components of an instream flow program are described, and current Texas methods for defining instream flow requirements are briefly reviewed. The chapter ends with three examples of current or recent instream flow efforts that use a number of the instream flow components, and research needs for the continued evolution of the science.

TRENDS AND PRINCIPLES OF INSTREAM FLOW SCIENCE

Trends in Instream Flow Science

In the United States, the National Environmental Policy Act (NEPA) of 1969 required federal planning activities to "create and maintain conditions under which man and nature can exist in productive harmony, and fulfill the social, economic, and other requirements of present and future generations…" This purpose of NEPA is reflected in many instream flow studies that seek balance among competing uses of water. Instream flow science began to develop in the years after NEPA in the late 1960s and 1970s, and continues to evolve today. Over these decades, four trends and seven principles mark the trajectory of instream flow science growth.

- *Hydrology and Hydraulics.* The convention of instream flow science is changing from developing a single, minimum flow or "flat-line" flow to a range of flows that account for seasonal and inter-annual variation, magnitude, timing, frequency, and rate of change (IFC, 2002; Poff et al., 1997; Postel and Richter, 2003). These hydrologic attributes translate into different levels of flow: subsistence flows, base flows, high flow pulses, and over bank flows (Figure 3-1). This range of flows is referred to as a **flow regime**.

Subsistence flow is the minimum streamflow needed during critical drought periods to maintain tolerable water quality conditions and to pro vide minimal aquatic habitat space for the survival of aquatic organisms. **Base flow** is the "normal" flow conditions found in a river in between storms, and base flows provide adequate habitat for the support of diverse, native aquatic communities and maintain ground water levels to support riparian vegetation. **High flow pulses** are short-duration, high flows within the stream channel that occur during or immediately following a storm event; they flush fine sediment deposits and waste products, restore normal water quality following prolonged low flows, and provide longitudinal connectivity for species movement along the river. Lastly, **overbank**

flow is an infrequent, high flow event that breaches riverbanks. Overbank flows can drastically restructure the channel and floodplain, recharge groundwater tables, deliver nutrients to riparian vegetation, and connect the channel with floodplain habitats that provide additional food for aquatic organisms. Increasingly, instream flow science promotes the inclusion of one or more of these flows in an instream flow study.

- *Biology.* The biological component of instream flows once focused on flow needs for one species (usually fish) and sometimes only one life stage of one species. Although single species remain the center of many instream flow evaluations, instream flow and riverine scientists now recognize and strive to account for multiple riverine ecosystem functions, sustained aquatic and riparian communities, and adequate habitat in instream flow programs (Calow and Petts, 1992, 1994).

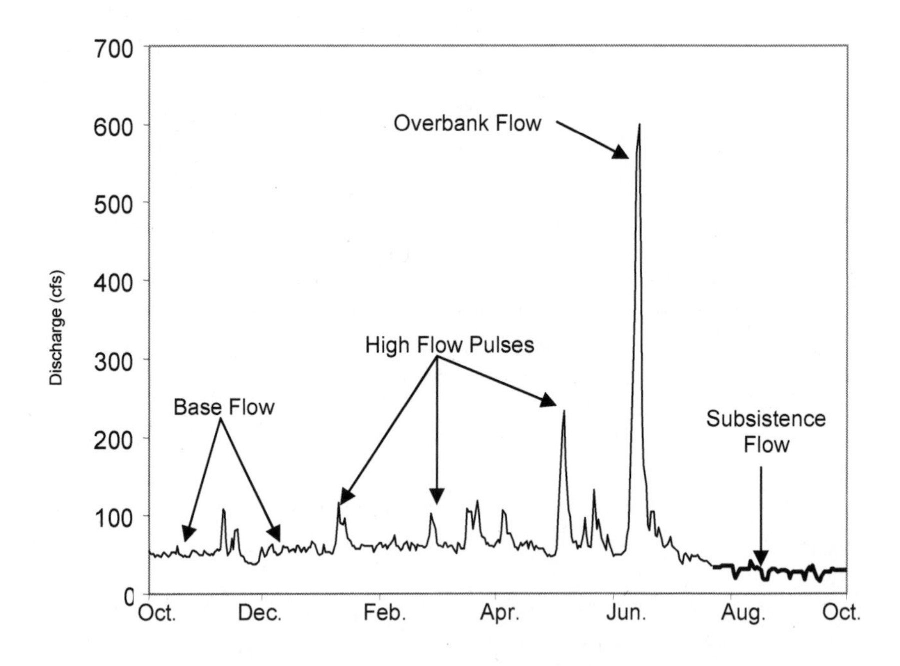

FIGURE 3-1 Daily streamflow hydrograph for Guadalupe River at Victoria, TX, with base flows, subsistence flows, high flow pulses, and overbank flows identified. SOURCE: Data from USGS Gage No. 08176500, water year 2000.

- *Geomorphology.* The stream channel used to be the spatial limit of instream flow work. Current goals, however, for state-of-the-art instream flow studies have expanded the spatial scope to include physical processes in the stream channel, riparian, and floodplain areas.

- *Disciplinary Focus.* In the 1960s and 1970s, hydrologists alone established the flow requirements primarily from hydrologic statistics (Orsborne and Allman, 1976). Increasingly, instream flow interdisciplinary teams have scientists from related fields of biology, geomorphology, water chemistry and quality, and water law and policy, as well as hydrology and hydraulics. The challenge of instream flow work is to develop an instream flow program that balances instream flow science(s), public values, and legal mandates. A multi-disciplinary team is best equipped to achieve and maintain this balance.

These four trends have made instream flow studies more comprehensive, but difficulties still exist for conducting instream flow studies. First, these trends result in studies that are more resource-intensive to conduct. Many times agencies simply do not have the staff, time, and monetary resources required to conduct this type of comprehensive instream flow study. And second, these more comprehensive instream flow considerations may further complicate the process of integrating results from disparate studies into a single flow recommendation. Still, the science is new and these obstacles, too, may be overcome with more research, information, and communication.

Principles of Instream Flow Science

There are several principles of effective instream flow programs included in the IFC's *Instream Flows for Riverine Resource Stewardship* (2002) and Postel and Richter (2003). The following list is adapted from these sources. These principles are reflected in the components of a state-of-the-art instream flow program and echo the four trends of instream flow science.

1) Preserve whole functioning ecosystems rather than focus on single species.
2) Mimic, to the extent possible, the natural flow regime, including seasonal and inter-annual variability (Figure 3-1).
3) Expand the spatial scope of instream flow studies beyond the river channel to include the riparian corridor and floodplain systems.

4) Conduct studies using an interdisciplinary approach. Instream flow studies need hydrologists, biologists, geomorphologists, and water quality experts all working together. Experts can come from academia, public, and private sectors.

5) Use reconnaissance information to guide choices from among a variety of tools and approaches for technical evaluations in particular river systems (see IFC, 2002 and Table 3-1).

6) Practice adaptive management, an approach for recommending adjustments to operational plans in the event that objectives are not being achieved (TPWD, TCEQ, and TWDB, 2003).

7) Involve stakeholders in the process.

The first three of these principles emphasize actions that should be conducted: preserve whole ecosystems, simulate the natural flow regime, and include floodplain and riparian zones in instream flow considerations. The last four principles offer means to accomplish the first three: take an inter- and multi-disciplinary approach; use a variety of tools; practice adaptive management; and involve stakeholders. Together, these seven principles reflect the scientific trends in instream flow science and provide the foundation for the components of state-of-the-art instream flow programs and studies.

COMPONENTS OF AN INSTREAM FLOW PROGRAM

Instream flow programs involve technical and non-technical components. Technical elements are the areas in which empirical or modeling evaluations are conducted: hydrology and hydraulics, biology, geomorphology and physical processes, water quality, and connectivity. Legal, regulatory, and public participation issues are some non-technical components of an instream flow program. Both technical and non-technical components are important in a state-of-the-art-instream flow program; otherwise, untenable situations can occur. For example, the most scientifically valid instream flow recommendation will not be implemented if it violates a permitting process, is out of compliance with water quality regulation, or lacks public support in the river basin. A successful instream flow recommendation will embody the seven principles and have clear goals, stakeholder involvement and support, technical evaluations, appropriate modeling approaches, integration of the various components, and adaptive management (IFC, 2002; Postel and Richter, 2003). Each component is briefly introduced below.

Clear Goals

Goals are statements of the activities or functions that instream flows are intended to support or achieve. Establishing clear management goals and objectives is an important component of any viable instream flow activity. River management personnel allocate stream resources among a variety of uses such as water supply, recreation, irrigation, and aquatic habitat protection. A lack of clear goals can create confusion as management agencies try to resolve competing demands or implement policy changes. Problems stemming from a lack of clarity in management objectives and authorities have been noted in several NRC reports of river systems across the United States, including the Colorado River (NRC, 1999), the Missouri River (NRC, 2002a), and the Upper Mississippi River (NRC, 2004b).

Ultimately, the act of setting goals for a program is a political action. In the case of instream flows, a heavy emphasis is on science, but the policy makers determine the parameters and focus of the instream flow program. Scientists subsequently carry out the technical evaluations accordant with these goals, and therefore need to play a strong role in setting the goals. The role of good science is to provide sound information that is useful in a forum for discussion by stakeholders and agency decision makers. Scientific input is critical to ensure that policy goals are consistent with scientific feasibility and that progress towards achieving the goals can be documented with measurable criteria.

Stakeholder Involvement

The IFC recognizes that public involvement and support are critical elements of instream flow programs (IFC, 2002). Several types of public involvement opportunities exist in an instream flow program: outreach and education, public hearings and meetings, and working groups. Stakeholder input can occur at several stages in an instream flow program. The public can participate in authorizing legislation, setting goals, and approving or commenting on instream flow recommendations. Public involvement can increase support for an instream flow program, and the benefits of public support for instream flow protection outweighs the costs of involving the public in the process (IFC, 2002; Postel and Richter, 2003).

Technical Evaluations

Technical evaluations are the sampling and modeling pieces of an instream flow study. These are often the heart of an instream flow study and consume the most resources. Technical evaluations of hydrology and hydraulics, biology, physical processes, and water quality involve empirical sampling or quantitative modeling. Connectivity involves the connections among and transfers between these aspects. In the best instream flow work, technical evaluations are closely aligned with the program and study goals. This alignment increases efficient uses of resources.

Hydrology and Hydraulics

Hydrology is potentially the most critical element of instream flow studies and has been considered the "master variable" because the biology, physical processes, and water quality components directly relate to it (Poff et al., 1997). Hydrology is used to assess hydraulic functions, water quality factors, channel maintenance and riparian forming processes, and physical habitat for target aquatic species. A flow regime encompasses the seasonality and periodicity of various types of flows, such as subsistence flows, base flows, high flow pulses, and overbank flows (Figure 3-1).

Hydrologic/hydraulic technical evaluations aim to understand and quantify the magnitude, frequency, timing and duration of subsistence, base, high pulse, and overbank flows; the degree to which the natural flow regime has been altered; descriptive aspects of the hydrologic system, such as location of springs, tributaries, and dams; and impacts of land and water use on the flow regime. Other examples of questions to be addressed in hydrologic/hydraulic technical evaluations are listed in Table 3-1.

Biology

Until recently, biology components in many instream flow prescriptions targeted one, or at best a few, important game or commercial species. Now, however, many new programs try to focus on whole riverine ecosystems. An instream flow biologic evaluation will assay fish species as well as invertebrates, amphibians, reptiles, birds, mammals, and riparian plants that are dependent on the river corridor for some portion of their life cycles. Depth, velocity, substrate, and/or instream cover constitute hydraulic habitat in aquatic systems, which is often emphasized in instream flow studies. Suitable hydraulic habitat is necessary, but it is not the only factor that affects the health of an aquatic ecosystem. Other factors that must also be considered include reproductive success of various species, disease outbreaks, predation, and competition for food.

Biological technical evaluations are often the main focus of instream flow studies, as habitat, life stages, or population dynamics are frequently the purpose of the study. Ecosystem processes can be difficult to measure or model, and biological sampling can be extensive in attempts to be comprehensive. To avoid wasting resources, biological technical evaluations should be tailored to the goals of the instream flow study and conducted in ways that are applicable to flow conditions. For example, aspects of biological instream flow sampling may refer to flow regime impacts on habitat, species of concern, or assemblages and life stages of species. Other sample questions for biological technical evaluations are listed in Table 3-1.

Geomorphology and Physical Processes

Physical processes form and maintain the shape of the stream channel and floodplain. The form of a river channel results from interactions among discharge, sediment supply, sediment size, channel width, depth, velocity, slope, and roughness of channel materials (Knighton, 1998; Leopold et al., 1964). The floodplain and riparian zone are also shaped by sediment transport and deposition. Stream channels react to changes in sediment dynamics and either degrade or aggrade along the longitudinal gradient in response to sediment load. Channel form provides the physical structure for habitat for aquatic organisms. Human modifications such as channelization and bank fortification impact the channel form and habitat. Instream flow technical evaluations of physical processes may document changes in channel structure, aquatic habitat composition, riparian vegetation, and other effects of physical processes in river systems. Other subjects of physical processes technical evaluations are listed in Table 3-1.

Water Quality

The primary assays of water quality in most instream flow studies are sediment and total suspended solids (TSS), nutrients, dissolved oxygen, and temperature. Temperature influences a variety of life history strategies of aquatic organisms and can impact fish migration, timing of spawning, length and success of egg incubation, growth rates, feeding behavior, or susceptibility to disease and parasites. Most aquatic organisms require moderate to high levels of dissolved oxygen, and the amount of dissolved oxygen affects biota in different ways, as different aquatic species can be highly tolerant or intolerant of low dissolved oxygen levels. Nutrient loadings to a stream can cause low levels of dissolved oxygen which can have deleterious effects on quantity and quality of habitat for macroinvertebrates and fish. Fine sediment and other suspended solids have well documented, negative effects on aquatic systems and represent a major source of degraded water quality in receiving waters throughout the United Sates (Waters, 1995).

Water quality issues are regulated at the Federal level by the Environmental Protection Agency and at state levels by agencies such as the Texas Commission on Environmental Quality (TCEQ). Water quality is not always included in instream flow programs because in many circumstances, the agency that administers water quality does not have jurisdiction over water quantity issues. However, water quality is relevant to instream flow efforts because water quality is highly dependent on water quantity and instream flows, and water quality technical evaluations should seek to highlight these connections. Sample questions that indicate connections between water quality and biological aspects and hydrologic/hydraulic aspects are listed in Table 3-1.

TABLE 3-1 Sample Questions to Guide Technical Evaluations

Technical Components of an Instream Flow Program	Suggested Questions for Technical Evaluations
Hydrology/Hydraulics	
Available data	Are the available hydrologic data sufficient for assessing the hydrologic conditions? Should monitoring be instituted where known deficiencies exist? Which statistical methods and tools (e.g., regionalization, record augmentation, disaggregation, etc.) can be utilized to develop needed data?
Flow regime	Are the available streamflow data sufficient to characterize annual and seasonal flow variability including the probability of floods or droughts? What is known about the magnitude, frequency, timing, and duration of base flows, subsistence flows, high flow pulses, and overbank flows? Should historical streamflow data be divided into pre- and post-development data sets? To what degree has the natural flow regime been altered?
Hydrologic system	Where are the major tributaries, major springs, dams, and diversions (including groundwater withdrawal) that influence the spatial pattern of flow? Is there longitudinal (upstream to downstream) connectivity in flow or are there major discontinuities (i.e., diversion dams), and if so where? What are the topographic and roughness

	conditions of the channel and floodplain? What are the stage-discharge relationships from nearby gaging stations? What are the statistical characteristics of streamflows?
Land and water use	What is known about the flow regime at key points in the watershed? What activities (e.g., trends in land use and surface water withdrawal, etc.) are influencing the flow regime and what are future projections for these trends? How do dam and reservoir operations impact flow regime and what are future projections for these operations?
Biology	
Available data	Are data from pre-project monitoring efforts available?
Flow regime	What is the importance of drought, flooding and intermediate flow conditions (flow variability) to habitat? What are the important connections to reservoirs or floodplains?
Species of concern	What species (fish, birds, mammals, invertebrates, aquatic plants or riparian vegetation) are of greatest concern from either ecological or socioeconomic standpoints? What times of year are most critical for these species?
Assemblages and life stages of species	Will modifications to current or naturalized flows protect habitat for the most flow-sensitive species or life-stages? Are flows sequenced to support life stages?
Physical Processes	
Geomorphic system	How do morphology and physical processes of the channel and floodplain vary spatially within the study area? Is the channel and floodplain system in dynamic equilibrium or disequilibrium? If the channel is a state of disequilibrium, what flow management scenarios could lead to a new equilibrium condition? Is the sediment input to each segment in equilibrium with the capacity of the channel to transport it through the segment? Is control of sediment input necessary?
Geomorphology and aquatic ecology links	How do physical habitat characteristics vary spatially? What physical features and processes provide key habitat for aquatic or riparian organisms of interest? What are current trends linking geomorphology and aquatic and riparian ecology? Can trends be reversed towards more naturalized conditions?
Land and water use	Has human activity and land use significantly altered the stream channel and floodplain morphology and processes? Do these alterations have a negative impact on key habitat? If so, what human activities are associated with this alteration? Are lateral channel migration, avul-

	sion, or meander cutoff processes important in this system, and have these processes been inhibited by flow alteration or other human activities?
Water Quality	
Available data	What is the present water quality status of the river segment? Are any of its designated uses impaired? If so, has a total maximum daily load (TMDL) study been done, and what are its results? Where are the wastewater discharge permit locations on the segment? What are their permitted flows? What proportion of the summer low flows in the river arises from upstream wastewater discharges? What is the current dissolved oxygen (DO) profile along the river? Has this changed appreciably in recent years? What is the stream temperature profile along the river? How does it change diurnally and seasonally? What is the total suspended solids concentration in the river? How does it change with discharge? How are water quality components affected by flow characteristics during the year and between different years?
Species of concern	What water quality components are of greatest concern to the target organisms, life stages, or riverine processes (DO, suspended sediment, temperature, chemical elements, nutrients)? Is the species distribution affected by water pollution (a typical consequence of polluted waters is a significant reduction in species diversity and an increase in pollutant tolerant species)?
Land and water use	Do land management activities affect water quality? If so, how do they affect riverine processes and organisms? Do opportunities (short- and long-term) exist to manage water quality-related factors in the watershed?
Spatial variability	Do water quality characteristics vary along the river, its tributaries, lakes, and estuaries (if any) throughout the watershed? If so, how do they change? Are these variations important?

Connectivity

Connectivity is "the flow, exchange, and pathways that move organisms, energy, and matter through river systems" (IFC, 2002). An instream flow evaluation should consider connections among hydrologic, biologic, geomorphic, and chemical aspects of instream flow. Examples of important connections are floodplain development processes, transfer of mass and energy from upstream to downstream positions, and vertical connections between surface and groundwater processes. Typical barriers to connectivity include dams, diversion structures, thermal effluents, organic loadings, toxic effluent discharges, and managed flow releases that can affect nutrient cycling, displacement of aquatic communities along the river continuum (Ward and Stanford, 1983), biodiversity, and environmental heterogeneity.

Conceptual Models

Often, technical evaluations are conducted independently of each other and the results subsequently combined into a single flow recommendation. A more efficient approach is to design the technical evaluations such that each sampling or modeling effort is tied directly to program goals; and the results of these evaluations are connected to aspects of the flow regime (Postel and Richter, 2003). These connections comprise a conceptual model. A conceptual model provides structure to integrate the disparate studies into a single flow recommendation and let individual scientists understand how all the pieces are intended to fit together. Also, these relationships will help focus efforts to meet the goals of the instream flow program. Table 3-2 is an example of a conceptual model that can help the instream flow team structure the process of integrating various technical aspects into a flow recommendation. Instream flow team scientists and others involved would begin to populate the cells in this matrix according to the river basin to be studied.

This approach represented by Table 3-2 treats flow as the master variable (Poff et al., 1997), but takes into account that water flow is one of multiple fluxes that impact ecological environments in rivers. The conceptual model of Table 3-2 includes other input and fluxes of sediment, nutrients, and organics in streams. For example, high rates of flow can impact other fluxes that erode banks, scour stream beds, and increase dissolved oxygen (DO). Ultimately, however, flow is the primary focus of instream flow recommendations and these other fluxes must be understood in terms of instream flow.

As simple as it appears in a table, many of these connections are difficult to make in practice. In fact, several, distinct conceptual models (like Table 3-2) could be constructed for the same river basin, depending on specialists involved and the goals of the program. One way approach to building a conceptual model is to pose and answer a series of descriptive questions that relate to the crux of the river basin instream flow issue(s). Samples of these questions are listed in Table 3-1, although this list is not exhaustive for all rivers in all regions. These questions can also be used to

TABLE 3-2 Relating Technical Components to the Flow Regime in an Instream Flow Study

Components of the Hydrologic Regime	Hydrology	Physical Processes	Biology	Water Quality
Subsistence Flow	Minimum streamflow needed to maintain tolerable water quality conditions and provide minimal aquatic habitat	Increase in deposition of fine particulate materials, notably organic material	Aquatic habitat is restricted	DO decreases, T increases
Base Flow	"Normal" flow conditions found in a river		Base conditions for aquatic habitat	
High Flow Pulse	A short-duration, high flow within the stream channel during or immediately after storm events	Flushing flows; connection to low-level off-channel water bodies; channel maintenance	Recruitment events for water-propagating species	Increasing levels of bacteria, TSS
Overbank Flow	An infrequent, high flow event that overtops the riverbanks	Floodplain construction and maintenance; connection to off-channel waterbodies; channel alterations; large woody debris recruitment and transport	High connectivity between aquatic and floodplain systems, yielding biotic exchanges between channel and floodplain	Increases in TSS and sediment loads

focus the instream flow study and ensure that only the most relevant information is empirically collected or quantitatively modeled.

Modeling Approaches

The complexities of riverine science are becoming better understood by natural resource professionals in part due to the emerging application of sophisticated assessment models. Many types of models can be and are applied to instream flow science. Hydrologic and hydraulic models, water quality models, sediment dynamics models, and biological habitat and life stages models are often presented as reliable approaches to derive flow recommendations (see IFC, 2002 for an exhaustive list of models).

Models have an important role in instream flow studies, but models must be chosen carefully to ensure that their input requirements are within the available resources and their output useful to derive a flow recommendation. The level of sophistication and the corresponding level of reliability of a model also have to be considered. Currently, a frequently used model in instream flow science is the Physical Habitat Simulation (PHABSIM) model. PHABSIM is a software model that quantifies hydraulic habitat attributes of selected species and life stages as a function of discharge. PHABSIM is part of the Instream Flow Incremental Methodology (IFIM), a modular decision support system for assessing potential flow management schemes. IFIM quantifies the relative amounts of total habitat available for selected aquatic species under proposed alternative flow regimes (see IFC, 2002). Despite their common applications, IFIM and PHABSIM have limits which may impact their applicability in Texas streams; they best fit small, clear streams with flagship aquatic species like trout or salmon and may not work well in blackwater or Coastal Plain systems.

Model selection requires consideration of the trade offs between model sophistication and wide-ranging application. Biological models, for example, range from holistic models that represent ecosystem processes to those that are species-specific. Correspondingly, instream flow biological models can range from low certainty to high certainty. Examples of the range of some of these models are presented in Table 3-3. The appropriate level of sophistication of biological estimation models for an instream flow study oftentimes lies somewhere in between these two extremes, depending on goals and river conditions on a basin-specific basis.

Integration

Integration in an instream flow program is the process of combining technical and non-technical input into a single flow recommendation. This step is an important step; unfortunately, state-of-the-art methods are not well documented in the current literature. Integration methods for technical evaluations are still being developed empirically. Two empirical examples are presented below. These examples illustrate two integrative ways to derive variable instream flow needs: a "building block" approach and a "percent-of-flow" approach. These approaches, of course, may not work for all rivers in all regions, but some aspects of the approaches are widely applicable.

The building block approach[1] builds a recommended instream flow hydrograph, or set of hydrographs, using key pieces of information developed during technical studies. For example, the findings of the technical biology studies may suggest that base flows of one level are needed during one season to maintain aquatic organisms, but base flows of different levels are needed in other seasons to enable fish movements up- and downstream. High flow pulses may be needed during specific times of the year to enable fish to access oxbows or floodplain areas for spawning or feeding. These specific flow needs, defined in terms of particular magnitudes of flow needed during specific months or seasons (or during certain years), can be used as building blocks to form an integrated instream flow hydrograph. High flow pulses and overbank flows are added on top of base and subsistence flows to construct the final recommendation. Different hydrographs may be prepared for different water years (dry, average, wet) to provide specific habitat needs or to facilitate various ecological processes. Figure 3-2 represents an example of an integrated hydrograph based upon the building blocks of subsistence flows, base flows, high flow pulses, and overbank flows.

The building block approach is particularly useful in river basins that have experienced considerable water development like dam construction. In such basins, the instream flow goals may be focused on re-building components of the hydrograph that have been altered considerably. In such instances, water managers can focus on restoring the key hydrograph components represented by the building blocks.

The second approach, "percent-of-flow" (Flannery et al., 2002; Figure 3-3) uses the technical studies to determine appropriate levels of allowable

---

[1] The building block approach or method was developed by Jacqueline King of the University of Cape Town, South Africa. For further information see King and Louw, 1998 and Tharme and King, 1999.

TABLE 3-3 Ordination of Some Basic Biological Assessment Methodologies from Holistic to Specific

	Holistic	←	→			Specific
	Higher Uncertainty	←	→			Lower Uncertainty
	Ecosystem Indicator	←	→			Metrics with Direct Response to Flow
Approach	Course Ecological Indicators	Assemblage Structure	Habitat Guilds	Species HSC	Population Models	Individual-based Models
Key Characteristics	Integrates many components/ processes; correlational	Integrates many components; correlational	Integrates many components; correlational	Individual or species; correlational	Dynamic simulation of aggregate response variable	Dynamic simulation of ecological mechanisms
Strengths (Benefits)	Rapid, cheap, Repeatable	←	→			Predictive with high resolution
Weaknesses (Limitations)	Ecological responses and mechanisms not specified	←	→			System site specific; expensive; time consuming
Settings where appropriate	Any	←	→			Those for which much information is available
Appropriate spatial scales	Intermediate to large	Intermediate to large	Intermediate	Small	Large	Small to large
Outputs	A single target value and correlations with flow	A set of values and correlation with flow	A set of values and correlation with flow	A series of values and correlation with flow	A series of simulated values at variable flows	A series of simulated values at variable flows
Examples/Applications	Multiple applications: water quality, watersheds, etc.	Channel-floodplain connectivity	Leonard and Orth, 1988	Many ISF programs	Long history but few formal applications to ISF	Jager et al., 1997, 2001; Railsback et al., 1999, 2002; others

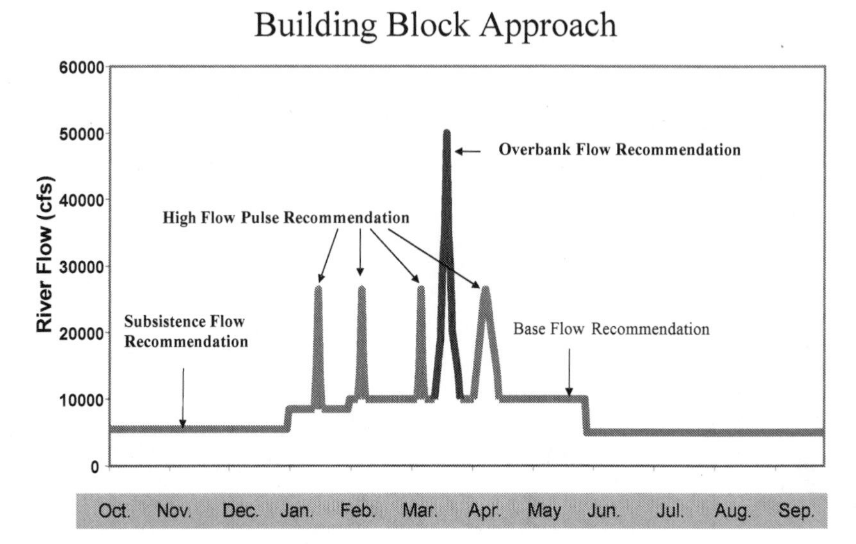

FIGURE 3-2 Example integrated hydrograph based on building blocks of subsistence, base, high flow pulses, and overbank flows.

flow depletion (typically expressed as percentages of the natural flow) during different water year types. This approach is particularly useful in river basins in which much of the natural flow volume and seasonal patterning remains and instream flow goals aim to mimic the natural ecosystem character.

Although integration of the technical pieces leads to the quantitative flow recommendation, the integration phase should also account for legal, institutional and/or socioeconomic issues that may influence the implementation of the instream flow recommendation. A number of formal analytical methods that might be applied to integrate social, economic, and legal considerations are available (see Kraft and Furlong, 2004). Stakeholder input and involvement are also important to provide insight to the local social and economic manifestations of the flow recommendation.

Adaptive Management

Adaptive management is widely recognized as a powerful approach to manage complex and dynamic situations (NRC, 2004c). Adaptive management is sometimes referred to as "learning by doing" and it is driven

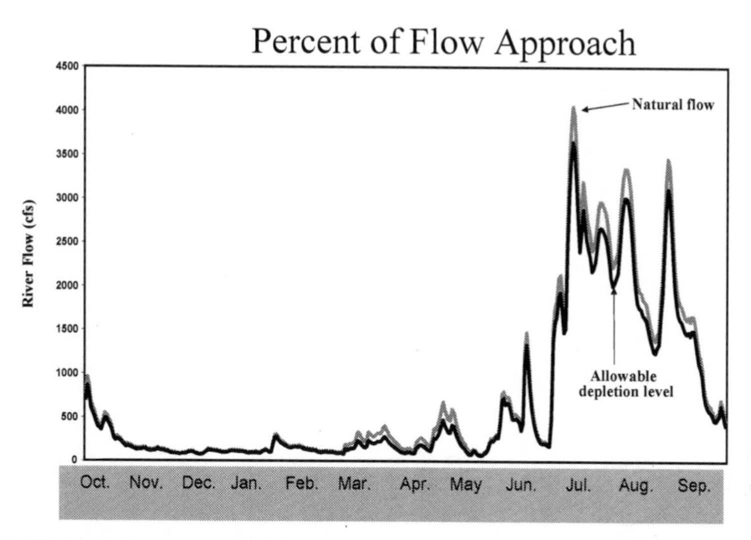

FIGURE 3-3 Example hydrograph based on the percent of flows approach for the Peace River in Florida.

by the goals of the program (Postel and Richter, 2003). There are five iterative steps to adaptive management in instream flow work: (1) develop goals; (2) develop or revise conceptual model; (3) develop or revise the flow prescription; (4) implement strategies for restoring flows; and (5) monitor and assess attainment of goals (Postel and Richter, 2003).

Ideally, instream flow programs are long-term enterprises that take several years to establish and additional years to incorporate the necessary study iteration and monitoring. Adaptive management is particularly useful in such studies, as it can test (and revise as necessary) the initial implementation of an instream flow program, assess ecological responses to new flow regimes, and add flexibility to the program and methods in the event that goals are not achieved. Therefore, a commitment to long- term monitoring, and anticipation that methods and flow recommendations may need revision over several years, are hallmarks of an adaptable instream flow program.

INSTREAM FLOW EXAMPLES

Many state instream flow programs have been in place for years. However, few of them provide much more than minimum levels of base flow protection and fewer still have validated flow prescriptions empirically. A state-of-the-art instream flow program takes time and resources to design and implement. Three examples of instream flow approaches are presented. The first example is a recount of the existing methods used in Texas to define instream flow requirements. The last two are more recent studies that highlight one or more state-of- the-art components. The Savannah River example (Georgia and South Carolina) shows the benefits of stakeholder involvement in the processes of developing goals and establishing instream flows. The Instream Flow Study of the Lower Colorado Basin (Texas) illustrates the utility of "critical flows" in determining instream flow recommendations.

Existing Methods for Defining Instream Flow Requirements in Texas

Texas currently has two hydrologic methods for defining instream flow requirements; one for water permitting (Lyons method) and the other for water planning (Consensus Criteria for Environmental Flow Needs, CCEFN). The Lyons method was developed by a fisheries biologist at the Texas Parks and Wildlife Department (TPWD), Barry W. Lyons (Bounds and Lyons, 1979). The approach uses percentages by month of daily-averaged flows (see IFC, 2002; Tennant, 1976) as the parameter that determines instream flows in Texas streams. For permitting, instream flows are 40 percent of the median monthly flows from October to February; and 60 percent of the monthly median flows from March to September. The 60 percent values were chosen to provide more protection during the critical spring and summer months. The 40 and 60 percent levels were determined using the wetted perimeter relationship of the river, i.e., the amount of river bed and banks that are wetted from stream flow. At 60 percent of monthly median flow, more than 80 percent of the river substrate was wetted, but below 40 percent of the monthly median flow, the percentage of wetted substrate began to drop off significantly as portions of the stream bed were exposed due to the low water conditions (Figure 3-4). These threshold levels have been applied to most rivers in Texas to determine existing instream flows for water permitting.

The second method that Texas uses to determine instream flows is the CCEFN, which is part of the Texas Guidelines for Regional Water Plan Development, produced by the Texas Water Development Board (TWDB, 2002b). These criteria are the result of collaboration among state agency scientists and engineers and local water resources representatives. CCEFN will be used in the second round of regional water planning in Texas, due to

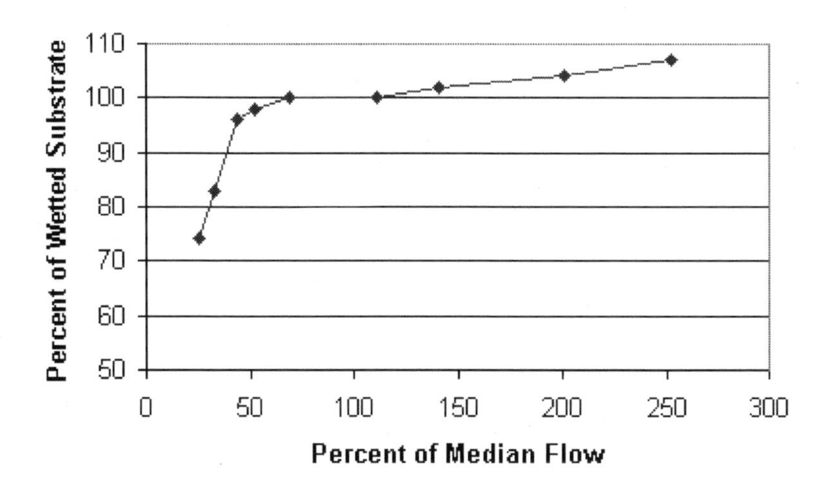

FIGURE 3-4 Variation of wetted substrate with streamflow on the Guadalupe River below Canyon Reservoir.
SOURCE: Data from Bounds and Lyons, 1979

be completed in 2007. All criteria are based on naturalized flows—the estimated flow that would have been present in a watercourse with no direct manmade impacts in the watershed. Criteria are defined in three zones for pass-through flows in reservoirs and for direct diversions from free-flowing streams and rivers (TWDB, 2002a). Whereas the Lyons method uses gage data as its flow value, CCEFN uses percentile values of the naturalized flow to determine direct diversion and pass-through flows. Unfortunately, this bifurcated approach to instream flow determination in Texas has created a system where the two methods produce different results for the same river (see Box 6-3 for further discussion).

Flow Recommendations for the Savannah River

The instream flow work in the Savannah River in Georgia and South Carolina began in 2002 and continues today. The U.S. Army Corps of Engineers (USACE) initiated a Comprehensive River Basin Plan to assess the degree to which various human needs and values for the Savannah were addressed through USACE water management, and whether changes in USACE dam operations might be warranted. With sponsorship from Georgia and South Carolina, the USACE worked with The Nature Conservancy (TNC) to facilitate a process for developing flow recommendations to protect and restore the river, floodplain, and estuary ecosystems in the lower Savannah River.

TNC organized an orientation meeting for stakeholders and interested parties. More than 60 scientists, water managers, and other representatives agreed on a one-year process to develop an initial flow recommendation that USACE could incorporate into its comprehensive plan for the river. The participants also identified specific scientists who should be involved in the process, as well as information sources thought to be useful in developing a flow recommendation.

After the orientation meeting, the University of Georgia's River Science and Policy Center produced a literature review and summary report (Meyer et al., 2003). The summary included statistical assessments of the available hydrologic data, a summary of the linkages between flow variations and the life cycles of numerous plants and animals, and a set of conceptual models of key hypotheses about flow-biota connections and human influences on key flow characteristics. These documents (Meyer et al., 2003) were circulated to more than 50 scientists identified during the orientation meeting who were invited to participate in a 3-day workshop to develop a flow recommendation for the Savannah River (Figure 3-5). Forty-seven scientists from more than 20 different state and federal agencies, academic institutions, and other entities participated in the flow recommendations workshop. During the workshop, they specified detailed flow requirements for a long list of target species and key ecosystem processes. The resulting flow recommendations differ among wet, average, and dry water years, and geographic location along the river.

The Savannah River project is on-going, and it is too soon to determine the degree to which the flow recommendations have achieved the goals of the project. Still, the process of developing goals and deriving flow recommendations used in the Savannah River project shows how stakeholders and scientists can collaborate successfully on instream flow studies. For river basins that seek strong stakeholder involvement in an instream flow project, the Savannah River project is a working example of how stakeholders advance the process.

Instream Flow Study of the Lower Colorado Basin

The instream flow study of Mosier and Ray (1992) is a landmark study because it was the first comprehensive instream flow study carried out on a Texas river. The Mosier and Ray (1992) study is instructive to examine the way in which hydrology, biology, geomorphology, and water quality were

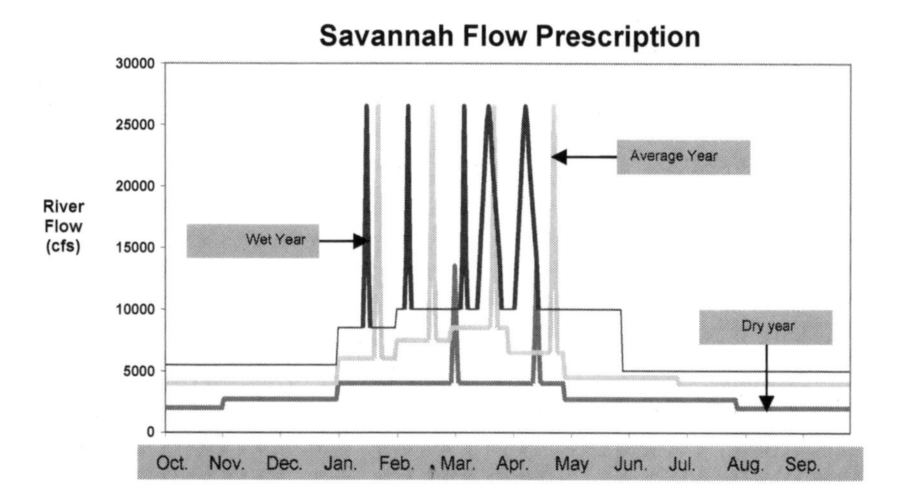

FIGURE 3-5 Flow recommendations for the Savannah River.
NOTE: This is only one possible translation of the flow recommendations. For each water year type, a number of high flow pulses of varying magnitudes is specified to occur within a particular time window.

drawn together to provide instream flow recommendations for the Colorado River below Austin, TX.

The instream flow study was undertaken in response to a condition mandated in the 1988 adjudication of the Lower Colorado River Authority's (LCRA) water rights in the river, and was carried out collaboratively by the LCRA and the TPWD. Upstream of Austin, the LCRA operates a sequence of six dams known as the Highland Lake reservoirs, and thus maintains significant control over flows in the lower river. From March to October, water is released from the Highland Lakes to supply water for rice irrigation along the Colorado River near the Gulf Coast. The release of irrigation supply water produces very large diurnal variations in discharge immediately downstream of Austin. During the winter months, irrigation water releases do not occur, and municipal wastewater discharges from the City of Austin are a significant part of the baseflow of the river immediately downstream of Austin.

In making instream flow recommendations, Mosier and Ray (1992) defined four types of flows:

- **Subsistence flow**—the flow needed to maintain water quality conditions, especially dissolved oxygen levels, considered adequate to support the native aquatic community. Mosier and Ray made specific recommendations for the Lower Colorado River using historical flow patterns and the QUAL-TX water quality model.

- **Target flow**—the flow regime that maximizes physical habitat complexity for the various components of the native aquatic community (see Figure 3-6). Hydraulic habitat analysis results in a schedule of monthly flows designed to optimize community diversity under conditions of normal rainfall. Under drought conditions, Mosier and Ray (1992) recommend reducing the discharge below the target flow but not below the subsistence flow.
- **Critical flow**—the flow distribution over time needed to support critical life history stages of certain components of the community, such as spawning and survival of fry.
- **Maintenance flow**—the flow conditions needed to scour the channel and prevent excessive siltation and macrophyte growth. Mosier and Ray (1992) offered the general recommendation that such flow pulses re needed but did not recommend a specific regime for them.

The study of Mosier and Ray (1992) was mainly focused on how the patterns of releases from the Highland Lake reservoir system could be optimized to support aquatic life in the downstream river. Its results were used as part of LCRA's Comprehensive Water Management Plan and resulting adjustments to its water permits for operating the Highland Lakes a were made by the Texas Natural Resources Conservation Commission (now TCEQ).

Many of the rivers included in the proposed instream flow studies (lower Sabine, Trinity, Brazos and Guadalupe) all have large upstream reservoir systems whose releases affect their flows in an analogous manner to the lower Colorado River. The instream flow study of Mosier and Ray (1992) is a valuable guide as to how similar studies could be undertaken in those rivers.

RESEARCH NEEDS FOR INSTREAM FLOW SCIENCE

Instream flow science continues to evolve in its philosophy and application. Research is critical to its evolution. Instream flow research has gained momentum over recent years, particularly in areas that focus on technical disciplines of aquatic biology, hydraulics, hydrology, and geomorphology, and emerging technologies for sampling the river environment. Multi-disciplinary instream flow studies that combine two or more of these fields are also more common (see the International Symposia on Ecohy-

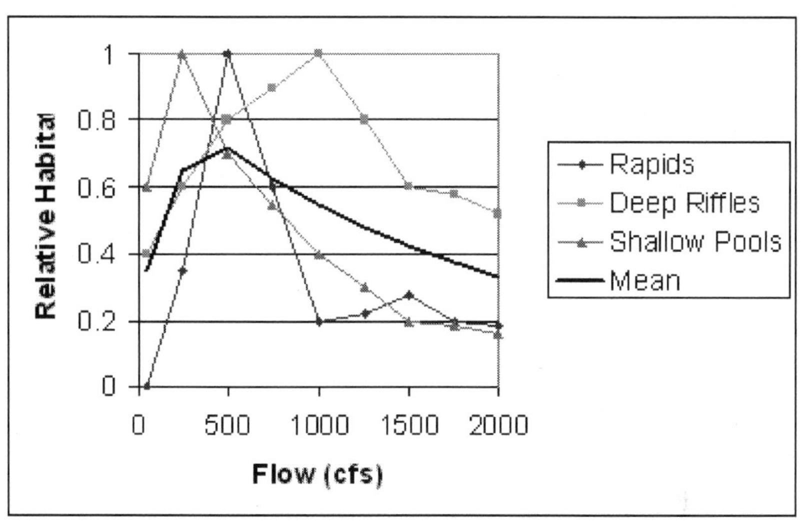

FIGURE 3-6 Habitat availability relative to a value of 1.0 at the target flow for the Colorado River at Bastrop. SOURCE: Mosier and Ray, 1992.
NOTE: Figure depicts three relative habitat curves, for rapids, deep riffles, and shallow pools, and the mean relative habitat curve formed by averaging over 10 habitat types, rather than the three depicted.

draulics[2]). With these advancements, major research needs and uncertainties still exist in the science of instream flows, especially with respect to integration, ecological indicators, and spatial scale.

In instream flow science, integration combines the different technical components into one recommendation or a set of flow recommendations. Integration is an important, complicated step in the instream flow process. Integration methods are being developed empirically. Anecdotal accounts indicate that instream flow integration has been done several ways, such as having scientists make the decisions, involving stakeholders in the process, using quantitative models, and combinations of all three. These different approaches have not been researched in terms of cost, timeliness, applicability, or accuracy. No conventional methods define the state-of-the-science for how integration is done and no evaluation of current options exists in the peer-reviewed literature. Furthermore, methods used to integrate results from disparate studies into a flow recommendation have not been well documented. In order for the state of instream flow science to advance in this area, integration methods will need to be established, re-

---

[2] Further information on the most recent International Symposium on Ecohydraulics can be found online at *http://www.tilesa.es/ecohydraulics/english/presenta.html.*

viewed, and refined. Therefore, more information, research, and documentation are needed about the process of reaching final flow recommendations to strengthen instream flow science.

Indicators are measurable quantities or variables that can be used to determine the degree that flow recommendations achieve the goals of the instream flow study or program. Indicators are important role in long-term instream flow monitoring and adaptive management. Ultimately, indicators guide informed policy decisions (NRC, 2000). Benefits of measurable indicators have been documented (GAO, 2004; NRC, 2000), along with the challenges associated with realizing those benefits, such as ensuring a sound indicator development process, obtaining sufficient data for reporting, coordinating data from multiple sources, and linking indicators to management programs and activities (GAO, 2004). Additional research is needed to develop criteria for ecological indicators (NRC, 2000) for use in instream flow studies.

The physical, chemical, and biological processes of a stream ecosystem operate at different spatial scales and are expressed in different spatial dimensions over daily, seasonal, annual, and longer time periods (TPWD, TCEQ, and TWDB, 2003; Ward, 1989). Instream flow requirements must accommodate these processes at their respective, multiple scales. Determining appropriate scale(s) for instream flow work is challenging because the scale(s) must be fine enough to conduct field sampling and coarse enough to apply to larger regions and be efficient in use of resources. The success of integration methods and ecological indicators is very closely linked to spatial scale in instream flow work. For example, integrating disparate study results from biology and geomorphology technical evaluations will be more effective if the separate studies are conducted at similar or comparable spatial scales. A single set of ecological indicators (specific to river basins) needs to be selected carefully to ensure right process or function at the right spatial scale. The difficulty, therefore, is determining the appropriate spatial scale for instream flow study design, selection of models and tools, and integration of study results (TPWD, TCEQ, and TWDB, 2003). Spatial and temporal scaling issues remain an important, viable research area for instream flow science.

SUMMARY

Instream flow is a simple concept with the difficult task of balancing competing uses for river water. Over the three decades of instream flow work in the United States, four trends have marked its evolution:

- from single, minimal flows to flow regimes;
- from a single-species focus to a focus on whole ecosystems;
- from the study of the stream channel to the study of riparian and floodplain areas, as well; and
- from a hydrology dominated field to an interdisciplinary field that includes hydrologists, biologists, lawyers, geomorphologists and water quality experts.

State-of-the-art instream flow programs will strive to preserve whole ecosystems, mimic natural flow regimes, include riparian and floodplain systems in addition to the stream channel, take an interdisciplinary approach, use a variety of tools and approaches in technical evaluations, practice adaptive management, and involve stakeholders. Instream flow programs will encompass technical evaluations in biology, hydrology and hydraulics, physical processes, water quality, connectivity, and non-technical aspects of stakeholder involvement and goal setting. Integrating technical evaluations into a flow recommendation is an important, challenging task with few well documented methods. Three examples of current or recent instream flow work are highlighted that use a number of these components and show how instream flow studies and programs work in Texas and across the country. Still, there are some major research needs and uncertainties in the science of instream flows, especially with respect to integration, ecological indicators, and spatial scale.

4

Evaluation of the Texas Instream Flow Programmatic Work Plan

The *Texas Instream Flow Studies: Programmatic Work Plan* (PWP; TPWD, TCEQ, and TWDB, 2002) lays out the rationale, background, and basic purposes of the Texas instream flow program and describes the process for conducting subbasin studies. The Texas instream flow program is described in the PWP and its companion document, *Texas Instream Flow Studies: Technical Overview* (TOD; TPWD, TCEQ, and TWDB, 2003). The TOD outlines the technical aspects of instream flow studies, including sampling methods. This chapter presents a brief overview of the PWP contents, recaps the strengths of the PWP, identifies areas for PWP improvement, and presents several suggestions for revisions and improvement. The TOD is evaluated in Chapter 5.

OVERVIEW OF PWP CONTENT

The PWP is a relatively brief (17 pages) document (TPWD, TCEQ, and TWDB, 2002). The portions most relevant to this report are abstracted here.

Agency Roles and Responsibilities

The Texas Commission on Environmental Quality (TCEQ) is the agency charged with implementing the constitution and laws of the state relating to water. Its responsibilities include jurisdiction over water and water rights and the state's water quality program. The Texas Parks and Wildlife Department (TPWD) has primary responsibility for protecting the state's fish and wildlife resources, and the Texas Water Development Board (TWDB) is responsible for water planning and financing for the needs of people and the environment. All three cooperating agencies are expected to participate in all aspects of instream flow studies, with one or more agencies assigned to take responsibility for coordination and planning of individual components of each study.

Legislative Mandate

Texas Senate Bill 2 directs the TPWD, TCEQ, and TWDB, in cooperation with other appropriate governmental agencies, to "jointly establish and continuously maintain an instream flow data collection and evaluation program." The agencies were further directed by Senate Bill 2 to "conduct studies and analyses to determine appropriate methodologies for determining flow conditions in the state's rivers and streams necessary to support a sound ecological environment." These study results "will be incorporated into future regional and state water plans, and will become essential data for conservation of fish and wildlife resources and consideration in the state water rights permitting process."

Priority Instream Flow Studies

The PWP identifies six river subbasins for priority study, which will be addressed in the following order during 2003-2010: Guadalupe River (lower subbasin), Brazos River (lower subbasin), San Antonio River (lower subbasin), Trinity River (middle subbasin), Sabine River (lower subbasin), and Brazos River (middle subbasin). Four additional basins are identified as candidates for a second tier of studies in the event that priorities change or supplementary resources are made available: Guadalupe River (upper subbasin), Neches River, Red River, and Sabine River (upper subbasin). The exact process for selecting the priority and second tier studies is not described in detail in the PWP; however, potential water development projects and water rights permitting issues are identified as important factors.

Scope of Studies

The PWP specifies that studies will include hydrology, biology, geomorphology, water quality, and connectivity, and that studies will be conducted using an interdisciplinary approach. The PWP notes the challenges of an inter-disciplinary approach:

> Recognizing the constraints of time and resources, it will not be possible to address each of these components in a systematic or

quantitative manner in each subbasin that is studied. However, each component should be evaluated and documented in the planning phases of each study for its applicability, feasibility, and importance to accuracy of models and study results.

In terms of spatial scale, the PWP indicates that an instream flow study is "largely a fish and wildlife resource evaluation of a river segment, sometimes a more comprehensive subbasin evaluation, and rarely a comprehensive evaluation of an entire basin."

Instream Flow Study Elements

The PWP presents a flowchart (see Figure 4-1) which depicts the intended sequencing of the work to be conducted in a study. An accompanying table in the PWP lists the tasks associated with each segment.

STRENGTHS OF AND OPPORTUNITIES TO IMPROVE THE PWP

Overall, the PWP presents an ambitious program with a sound, skeletal foundation for a successful instream flow program. The agencies are com mended for identifying the need to evaluate the primary components of river systems. Still, the PWP offers opportunities for improvement to strengthen its programmatic structure. With the improvements suggested in this report, the PWP should provide the architecture necessary for Texas to build a successful instream flow program.

Strengths of the PWP

The PWP has several strengths. First and foremost, the PWP presents an approach that conforms to the best practices for instream flows as defined by instream flow experts. As part of this approach, the Texas agencies have identified the important and relevant elements of an instream flow study. Chapter 3 of this report identifies seven principles of state-of-the-science instream flow programs:

- Preserve whole functioning ecosystems
- Mimic, to the extent possible, a natural flow regime

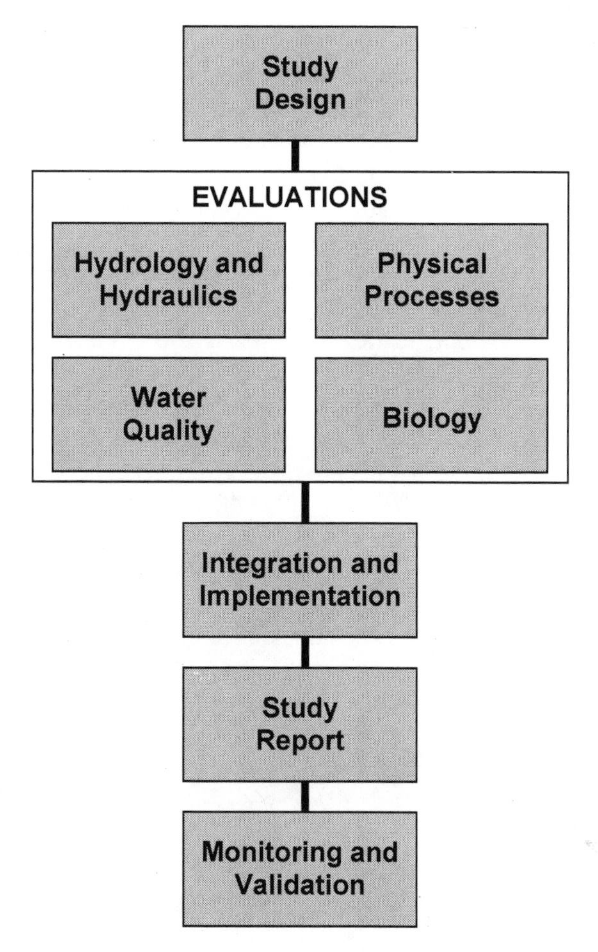

FIGURE 4-1 Flowchart of instream flow study elements.
SOURCE: Adapted from the PWP (TPWD, TCEQ, and TWDB, 2002).

- Include the riparian corridor and floodplain in the spatial scope of the study
- Conduct studies using an interdisciplinary approach
- Use a variety of tools and approaches appropriate for particular rivers
- Practice adaptive management.
- Involve stakeholders in all aspects

To the credit of the Texas agencies, the PWP includes all of these characteristics to some degree. The PWP also specifies and stresses the use of a multi-disciplinary tack that includes hydrology, physical processes, water quality and biology input. Finally, the PWP is very clear in identifying the priority study sites, outlining the roles of the state agencies, and emphasizing the importance of coordination among state agencies and other interests in conducting instream flow studies.

Opportunities to Improve the PWP

Several areas of the PWP need improvement. Two aspects of the PWP need immediate attention and improvement to validate the instream flow program presented in the PWP. First, the PWP needs to outline a plan to create a unified program with state-wide comparability that accommodates studies tailored to local conditions. Second, the PWP needs clearly articulated goals. These two aspects are major areas for PWP improvement. Other aspects of the PWP that need revision or clarification are the PWP flowchart, use of existing and reconnaissance data in the detailed technical evaluations, scaling issues, monitoring and validation, adaptive management, and stakeholder involvement.

State-wide Comparability with Studies Tailored to Local Conditions

Texas Senate Bill 2 directed the agencies to develop and maintain "an instream flow data collection and evaluation program" and the PWP generally refers to the instream flow effort as a program. It is assumed that the Texas instream flow program is intended to be more than a collection of individual studies. The challenge, therefore, is to construct an instream flow program with two levels of oversight: one to provide consistency at the state-wide level and one to accommodate individual differences at the subbasin level. The Texas agencies did a commendable job in identifying these two programmatic levels; however, the PWP does not discuss the connections between these two levels to work as a single, coherent program.

The state-level oversight should provide the structure to compare and, perhaps to some degree, integrate findings from subbasin-level technical evaluations. The state-level structure should also provide some consistency across instream flow studies that are tailored to the local, subbasin conditions. A consistent approach across basins promotes the efficient use of resources. For example, lessons learned from early studies can be applied to subsequent instream flow studies in different subbasins. If instream flow recommendations from the earliest studies do not appear to be having the desired effects, mid-course corrections (via adaptive management) might be possible for other priority studies. The state-level program should also allow results from studies in one or more of the priority river subbasins to inform management decisions in non-priority rivers. This information may be particularly relevant to the Texas program, as TCEQ has classified 225 segments on Texas streams and rivers for water quality purposes, but only eight of these 225 segments have been identified as priority areas for the instream flow program.

There are some general similarities among the rivers in Texas; they have low to medium gradients with relatively warm water. The state-level structure will cater to these similarities. However, there are also important differences across Texas river basins. Rainfall varies across Texas, and rivers in different parts of the state may have various levels of dependence on springs and other groundwater sources. Rivers across the state respond differently to human activities such as urbanization, wastewater return flows, and the existence of dams. Finally, there is significant biological diversity across the state. All of these factors support a second level of oversight in the Texas instream flow program that promotes studies that are designed based on the specific characteristics of the study subbasin.

The simultaneous need for statewide consistency and individually tailored studies may present a dilemma. Fortunately, there are demonstrated ways to address this issue. For example, the U.S. Geological Survey National Water Quality Assessment (NAWQA) Program[1] combines national consistency and local flexibility. NAWQA uses information collected in selected river basins across the country to address local, regional, and national water quality issues (for reviews on NAWQA, see NRC 1990, 2002b). NAWQA shows how environmental monitoring can be conducted successfully across many federal, state, and local agencies (NRC, 2002b). Attributes of the program that help it achieve national consistency include (1) clearly focused goals; (2) well documented methods and approaches; (3) site selection and sampling and analysis protocols that were designed to produce data and information that can be combined and interpreted in a broad context; and (4) national oversight and quality assurance, including review of individual study plans relevant to eventual application to national issues. While the Texas instream flow program is very different from NAWQA, these four attributes of NAWQA can be useful guidance to the Texas agencies in the process of articulating the two-level structure of the Texas instream flow program.

---

[1] For further information on the USGS NAWQA program see *http://water.usgs.gov/nawqa/*.

The PWP makes clear that all studies will be multidisciplinary, will follow similar steps, and will include plans for on-going monitoring and validation. The priority studies will be conducted over a period of at least ten years and a large number of people across the state are likely to be involved, including personnel from the state agencies, river basin authorities, academia, and private-sector consultants. Actions that will strengthen connections between the state-level program and subbasin-level studies include (1) extensive documentation of rationale, methods and approaches chosen for the technical evaluations conducted in each river basin; (2) documentation of the procedures used to integrate the results of individual disciplinary studies into an instream flow recommendation; and (3) continued oversight of the entire process by the state agencies and peer review. These three actions will provide the needed information and data structure to compare methods and results from different subbasin studies; integrate findings, as appropriate, from different subbasins; and share important instream flow study information across the life of the instream flow program and all of the state, academic and private sector personnel who will be involved in the program.

Programmatic and Basin-Specific Goals

Establishing unambiguous management goals and objectives is an important component, perhaps the most important component, of any viable instream flow program (see Goals section, Chapter 3). The PWP (page 2) contains two broad goals for the program in the statement: "the goal of an instream flow study is to determine an appropriate flow regime (quantity and timing of water in a stream or river) that conserves fish and wildlife resources while providing sustained benefits for other human uses of water resources." Unfortunately, sometimes the goal of conserving fish and wildlife may conflict with the goal of providing human uses of water. Thus, the trade-offs inherent in these two broad goals, neither of which may be able to be fully met, may present difficulties for instream flow management.

In addition to clear, state-wide programmatic goals, each individual river basin study will also need goals and objectives that are tailored specifically to that particular subbasin. The study flowchart presented in the PWP includes as part of study design the task, "develop objectives and study plan specific to subbasin" but no guidance is provided about the nature of these objectives or how they are determined. Because the goals for the subbasins are likely to reflect the wide range of interests and conditions of those basins, the PWP cannot, and appropriately, does not, dictate what the subbasin goals should be. Still, the PWP needs to mention that sweeping goals at the state, programmatic level need to frame the site-specific goals that will guide technical evaluations at the subbasin level.

One way to approach these two levels of goals is to have a state-wide goal for the instream flow program and subbasin goals that nest within that goal. One state-wide goal is stated in the Senate Bill 2 language: to conduct studies to support a "sound ecological environment" in Texas rivers. This is a clearly stated goal; however, neither the Bill nor the PWP defines the phrase "sound ecological environment," which has left its meaning open to interpretation. During public meetings with stakeholders in Texas over the course of this study, stakeholders presented widely different interpretations of a "sound ecological environment," from the preservation of natural biodiversity to industrial, commercial and recreational uses of rivers. The stakeholder comments underscore the import of establishing a single, statewide definition for this term.

Admittedly, developing these goals statements will not be a simple process. There are several options available to define and realize a "sound ecological environment" in Texas rivers. One option is to invite stakeholders into the process and define the goals by consensus (Postel and Richter, 2003). Another option has roots in the PWP which contains a strong statement about high quality, intact ecosystems in Texas:

> A high quality, natural environment is essential for conserving the quality of life Texans, future generations of Texans, and visitors to this state enjoy. Intact and functioning ecosystems are also critical for maintaining a strong state economy. Healthy aquatic systems that maintain biological integrity are essential to conserve the state's natural biodiversity, as well as support tourism, recreational pursuits, commercial and recreational fisheries, and a myriad of other industries.

This description of high quality aquatic ecosystems captures some important aspects of what a sound ecological environment could be (i.e., intact, functioning ecosystems, biodiversity, biological integrity, etc.). If Texas intends to use this description as proxy for a sound ecological environment, that intention should be stated explicitly. An inclusive approach like this statement is encouraged at the state level; however, this statement in the PWP would be even more useful if it had a stronger quantitative description that could be matched with measurable metrics.

Metrics measure progress towards achieving the selected management goal. Examples of metrics are number or abundance of some species, fish or macroinvertebrate populations, range of hydraulic habitat, etc. These and other metrics are well documented in the TOD. Instream flow management goals and metrics could be tied to the components of the flow regime (i.e., base flows, subsistence flows, high pulse flows, and overbank flows) to strengthen connections among the studies and between the studies and the instream flow recommendation(s). Regardless of whether the policy goal is set by stakeholders, legislation, or agency decision makers, clarification of the meaning of the phrase "sound ecological environment" is essential. Those tasked with conducting instream flow studies or implementing the recommendations that come out of the studies will need an unambiguous understanding of the term in order to design studies to comply with the goal of achieving sound riverine environments in Texas. Defining this term is primarily a policy decision, but this decision should be informed by scientific advice on alternative definitions and on metrics to measure progress toward the goals.

Instream Flow Studies Flowchart

A main strength of the PWP flowchart is its simplicity. The PWP instream flow flowchart (Figure 4-1) presents most of the important elements of an instream flow study and does it in a simple, straightforward fashion. Attempting to diagram a complex undertaking such as an instream flow study can be difficult because input from multiple steps must be considered at the same time. For example, the results of the hydrologic, biological, water quality and physical processes evaluations must be interpreted together to develop instream flow recommendations and each of these discipline evaluations is typically made up of a collection of separate studies. Thus, a diagrammatic representation can get quite complicated with multiple inputs and feedback loops (see for example Bovee, 1998). This complexity may be appropriate for a strictly technical audience, but those who do not have a technical background may not be well informed by an overly complicated diagram. Thus, there are advantages to a relatively simple and straightforward flowchart, and the Texas agencies are commended for presenting a complicated process in a diagram that is easy to understand. The PWP flowchart acknowledges the need for different disciplines and integration of results, and if applied as presented, the flowchart should promote a consistent approach to instream flow studies across the state. Its straightforward approach can simplify a complex process to non-technical stakeholders.

The potential problem with this streamlined approach is that connections between the presented steps may not be easy to understand because very step cannot be detailed. Supporting documents are critical to provide necessary detail and show linkages between and among the steps. In the text describing the PWP flowchart, several important steps do not get the emphasis and degree of description that are needed. These steps include: (1) establishing goals that are as clear and measurable as possible, (2) providing a process to incorporate existing information and reconnaissance studies into the design of the technical evaluations, (3) providing information at the study design step to guide choices about spatial scales for the technical evaluations, (4) establishing a process for integrating scientific results into an instream flow recommendation, (5) considering factors that may affect implementation of the recommendations, and (6) selecting indicators for long-term monitoring.

In order to assure that these steps get the necessary consideration in an instream flow study, the PWP flowchart and supporting text should be revised as follows. PWP revisions should give specific attention to goals; include a two-step process whereby existing and reconnaissance data are collected (first step) and used to design the detailed technical evaluations (second step); specify spatial scales during the study design of the technical evaluations; clarify the process for integrating information; consider implementation issues; and include more information about the use of indicators and monitoring. Also, while a study report is reasonable at the end of the process, a report should be produced after the conclusion of the technical evaluations and prior to implementing the flow recommendations. These suggested changes are presented diagrammatically in the revised flowchart presented in Figure 4-2. This flowchart, which is still relatively simple and straightforward, is based on the flowchart in the PWP (Figure 4-1) with the addition or reordering of major steps and tasks that are important to successful instream flow studies based on experiences in other places (IFC, 2002; Postel and Richter, 2003).

Use of Existing Information and Reconnaissance Studies

Among the tasks to be completed during the design of an instream flow study, as specified by the PWP, are compiling and evaluating existing information and field reconnaissance. The description or plan of how to incorporate these preliminary assessments in subsequent steps of the study is either weak or missing in the PWP and TOD. This is a difficult step in any instream flow program or study. The Texas agencies have laid the groundwork in the PWP, but the PWP needs further description to highlight the plan to realize this important step.

The Texas instream flow program can be viewed as a two-step process. The first step is the collection of existing and reconnaissance data. This step includes any initial studies that are needed to describe the most basic

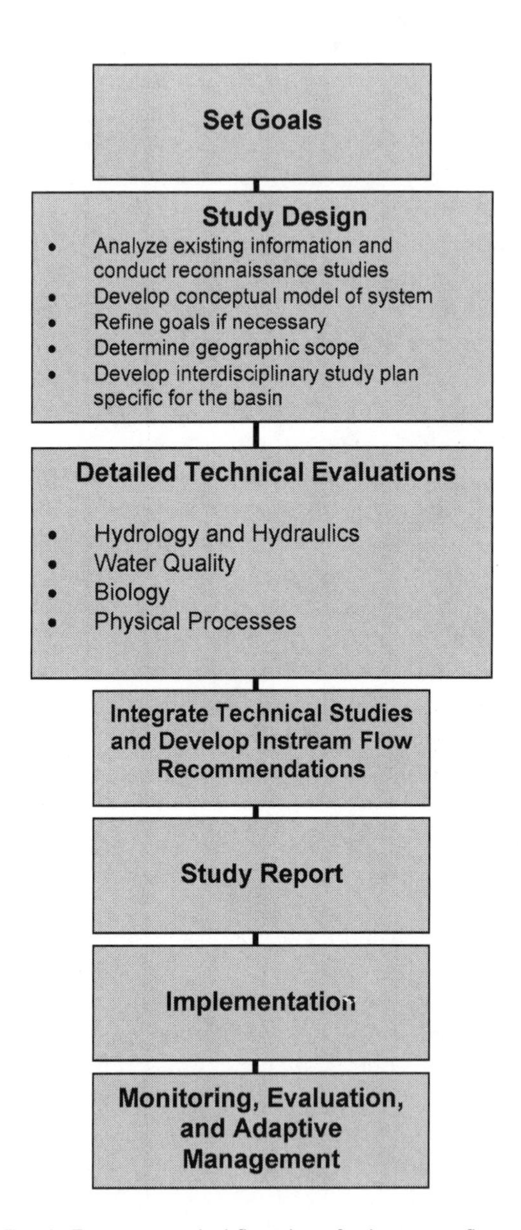

FIGURE 4-2 Recommended flowchart for instream flow studies.

aspects of the riverine system. The second step is the conduct of the detailed technical evaluations focused on hydrology and hydraulics, physical processes, biology, or water quality. Existing information and reconnaissance-level studies compiled in the first step should be used to design detailed technical evaluations, the second step. Information from the first step would be compiled into a conceptual model of the river system (see Chapter 3) to ascertain what is and what is not understood about the river system, and what additional, detailed information needs to be collected or modeled. A conceptual model can focus subsequent detailed technical evaluations in quantitative terms. A series of questions and answers can provide enough information to develop a conceptual model of a river system that includes the physical, biological, and water quality characteristics of the study area.

Examples from the Savannah River project include the kinds of questions that might be formed from the conceptual model (Meyer et al., 2003):

- What flow in March through May is needed to provide adequate larval drift for striped bass?
- What flow in January through April is needed to provide floodplain access for fish?
- What flow is needed every 5 years to form pool-riffle habitats in the stream channel?

Developing the conceptual model will marshal the expertise of every member of the multi-disciplinary instream flow team. Even still, posing the right questions and then setting out to answer them is a challenging exercise. A wide variety of technical tools and methods can be used to model, simulate, or quantify processes to answer such questions. Tools, methods, and models should be selected carefully to be as tailored to the study subbasin as possible to investigate the status of the physical, chemical, and biological characteristics of the river system under study. Furthermore, the selection and rationale for certain methods should be well documented for future reference. Once developed, a conceptual model can highlight missing data or gaps in understanding of certain important components. Recognizing these gaps is a useful outcome of the conceptual model. The PWP and the TOD need better explanation of the process whereby existing information and reconnaissance studies will be used to guide the detailed technical evaluations of hydrology, physical processes, biology, and water quality.

Designing the technical evaluations from existing and reconnaissance data is an involved, but important step in the instream flow process. When done correctly, the detailed technical evaluations will be aligned with program and subbasin goals and each other for a more streamlined integration process. When done poorly, the resource-intensive detailed technical evaluations can waste resources on sampling and modeling efforts that do not relate to program goals or advance progress towards a flow recommendation.

Scaling Issues

Spatial and temporal scaling issues remain at the forefront of research needs in instream flow work. Much uncertainty surrounds approaches to correctly scale instream flow empirical studies and applications. Spatial scale compatibility is critical at the point where disparate technical studies are integrated into a flow recommendation. In one view, the Texas agencies have considered spatial scale in instream flow studies very well. The lengths of the main stem river reaches in the six priority study segments (see Table 4-1) range from 137 to 272 river miles. The boundaries of these instream flow study reaches largely coincide with the boundaries of the water quality management segments established by the TCEQ as part of its Texas Water Quality Standards. Thus, water quality and instream flows are being analyzed using comparable spatial units. This is a strong point because the agencies have had experience working at this scale for the water quality program. It is also a benefit for future integration of the results of the instream flow and water quality programs, should the agencies choose to take advantage of that opportunity.

In another view, however, the PWP and TOD are not very clear on the selection of river reaches and segments for study. There is some guidance in the TOD about selection of representative reaches for hydrologic studies, but it is not clear that this selection process will result in study areas that are equally useful for the physical processes, water quality and biological components of the study. Agency personnel have extensive experience with Texas rivers and may have addressed these issues in other studies. If so, these studies can and should be referenced. The PWP and the TOD will be strengthened by addressing the issues of scale and comparability of studies conducted within the different disciplines. Ensuring that the different technical evaluations are conducted at commensurate spatial and temporal scales appropriate to derive an instream flow recommendation is the key scaling issue for the Texas instream flow program.

TABLE 4-1 Lengths of Main-Stem River Reaches in Priority Instream Flow Subbasins

River Reach	TCEQ Segment Number	Length (miles)
Lower Sabine	502 and 503	137
Middle Trinity	804	160
Lower Brazos	1202	199
Middle Brazos	1242	183
Lower Guadalupe	1803, 1804	272
Lower San Antonio	1901	153
Total	8	1,104

SOURCE: Data from TNRCC, 2000.

Monitoring and Validation

When water managers begin implementing an instream flow recommendation, it will be very important to monitor the degree to which instream flow goals are being met. This serves at least two purposes. First, if monitoring results suggest that the instream flow goals are not being met, it could provoke water managers and scientists to modify the instream flow recommendations. Second, if ecosystem benefits associated with implementation of instream flows can be documented, that documentation will help build societal and scientific support for the instream flow program.

The PWP recognizes the need for monitoring and validation components in the instream flow study process and the TOD accurately discusses a number of purposes served by long-term monitoring. The PWP and the TOD note that monitoring ecosystem conditions during implementation of the flow recommendations can validate the results of modeling conducted during the technical studies and gauge whether instream goals are being attained. The PWP and TOD do not, however, provide any guidance on the selection of components that are to be monitored.

Because of the importance of monitoring to program success, the PWP should specify that each study plan develop a suite of measurable ecosystem indicators that are responsive to instream flows and can be tracked to measure ecosystem conditions during the study and after implementation of instream flow recommendations. The instream flow study plan should explicitly identify the techniques to be used in monitoring the indicators, and frequency, locations, and timing of measurements. Indicators should be related directly to the goals of the subbasin study. Finally, the selection of indicators should be determined in the study design phase.

Adaptive Management

Adaptive management is recognized as a powerful approach to management in complex situations (NRC, 2004c). An adaptive management approach is encouraged to be used in the Texas instream flow program to account for mid-course corrections and respond to long-term monitoring results.

The PWP authors are commended for recognizing monitoring as necessary for adaptive management practices, but the PWP omits certain important points such as (1) specific assessment of instream flow recommendations in meeting target resource objectives; (2) specific description of a conceptual model (or how the different technical pieces fit together); and (3) evaluation of the overall implementation of the instream flow process, ecological models, tools and analyses employed. These are important elements of an adaptive management approach and should be included in any revisions made to the PWP. Further, the PWP is not clear about how management agencies might respond in circumstances when monitoring results suggest problems with the models or techniques used, selection of indicators, or shortcomings in attaining instream flow goals.

Stakeholder Involvement

Stakeholder involvement is included as one of the important principles for riverine resource stewardship by the Instream Flow Council (IFC, 2002). Stakeholder involvement at the goal-setting step is particularly important because of the potential for conflict among competing uses of water. In Texas, stakeholders are vested in and knowledgeable about instream flow issues. Based on stakeholder input at the committee open meetings in Austin and San Antonio, Texan stakeholders, if given the opportunity, could contribute to the instream flow process in significant, valuable ways. However, a vested stakeholder contingent does not equal a contingent in agreement. To the contrary, stakeholders rarely agree on how water should be used with respect to instream flows. Municipal demands, agricultural use, recreational interests, threatened or endangered species, and water-related regulations will all have to be taken into consideration. Since state agencies, stakeholders, and civic groups likely will disagree, it is important to allot adequate time to address the range of issues in setting instream flow goals and to have a pre-determined process to set goals if compromise cannot be reached.

The Texas instream flow documents indicate that an early step in conducting instream flow studies will be to identify stakeholders and potential cooperators. Stakeholders and cooperators are rather broadly defined in the TOD as federal agencies, river basin authorities, the academic community, environmental groups, recreational groups, and other interest groups. The TOD says that a stakeholder process will be developed but neither the PWP nor the TOD specify how stakeholder interests will affect study objectives or study design. The PWP will be improved if the role and degree of stakeholder involvement is clarified.

SUMMARY AND RECOMMENDATIONS

The PWP is a relatively brief document that describes the programmatic aspects of the Texas instream flow program. It lays out agency roles and responsibilities, its legislative mandate, priority instream flow studies in Texas, the scope of instream flow studies, and instream flow study elements. The strengths of the PWP include clearly articulated legislative mandate, identification of the priority studies, and general roles of the state agencies. The PWP presents an instream flow approach for Texas that is consistent with current thinking on instream flow best practices. It incorporates important and relevant elements of an instream flow study through a multidisciplinary approach that includes hydrology, physical processes, water quality, and biology.

The PWP also presents opportunities for improvement. Two major areas that need attention are (1) an explanation of an instream flow program that allows individual studies to be tailored to the study subbasin and consistency and management at the state level; and (2) articulation of clear goals. Other aspects of the PWP also need revision or clarification. The PWP needs to emphasize a two-phase process where existing and reconnaissance data are collected (first step) and used to design the detailed technical evaluations (second step). The PWP needs a clearer description or plan as to how existing and field reconnaissance informs the detailed technical evaluations. Additional emphasis is also needed on setting subbasin goals and explaining how results from the detailed technical evaluations will be integrated to derive a flow recommendation. Key aspects of spatial scale issues need further clarification, as well. Different technical evaluations need to be designed and conducted at spatial and temporal scales commensurate with each other and at an appropriate scale to derive an instream flow recommendation. The PWP mentions the value of monitoring and validation, and needs to identify indicators to be able to quantify progress through monitoring and validation activities. Adaptive management is briefly discussed in the PWP, but more detailed information is needed about the (1) specific assessment of instream flow recommendations in meeting target resource objectives; (2) specific description of a conceptual model; and (3) evaluation of the overall execution of the instream flow processes, models, and analyses employed.

Therefore, several recommendations for the PWP include:

1) A clear definition of the phrase "sound ecological environment" needs to be provided to supply context for instream flows in Texas.

2) The PWP should present a state-wide context for individual sub-basin studies. This can be accomplished with two levels of oversight: one at the state level for management and program consistency and one at the subbasin level for goals and approaches that are tailored to the specific needs of the study basin.

3) The PWP should present clear and specific goals for the state-wide instream flow program and recognize the need to develop individual sub-basin goals that nest within the state-wide instream flow programmatic goal(s).

4) The PWP flowchart for instream flow studies should be revised to include several important steps in planning and conducting an instream flow study as suggested in Figure 4-2.

5) The PWP and the TOD should describe how existing information and reconnaissance studies will be used to guide the detailed technical evaluations of hydrology, physical processes, biology, and water quality.

6) A suite of measurable, ecological indicators should be established for the state-wide program and each basin-specific study; the indicators should be responsive to instream flows. These indicators can be used in adaptive management, monitoring and validation activities to measure progress towards achieving a sound ecological environment in Texas rivers.

7) The PWP or TOD should provide information about how adaptive management will be implemented for the program as a whole and for individual river basins.

8) The PWP should provide additional information about the type and degree of stakeholder involvement in the instream flow studies.

5

Evaluation of the Texas Instream Flow Technical Overview Document

The Technical Overview Document (TOD; TPWD, TCEQ, and TWDB, 2003) outlines the methodological aspects of conducting instream flow studies in Texas rivers. The act of drafting this document is acknowledged as formidable because it must simultaneously provide (1) methods that are specific enough to guide technical evaluations, and (2) guidance that is broad enough to be applicable in individual subbasins across the different river systems in Texas. The Texas agencies faced a dilemma in writing the TOD because uniform approaches towards technical methods will be of little value to Texas with its wide range of riverine conditions, but the TOD cannot possibly make methodological prescriptions for every river system in the state. Therefore, the difficult task of preparing the TOD involves finding middle ground between these two options.

This chapter reviews and comments on the technical sections of the TOD and provides recommendations for its improvement. The TOD was evaluated for technical accuracy in the context of the instream flow program. The review of the TOD begins with a brief summary and description of the document's contents. Subsequently, a section on the overall findings of the TOD is presented, followed by individual evaluations of the subsections of hydrology and hydraulics; biology; physical processes; water quality; and integration and interpretation in the order they are presented in the original TOD. Implementation aspects are discussed in Chapter 6.

OVERVIEW OF TOD CONTENT

The TOD describes the methods to be used to collect, analyze, and integrate technical information among hydrologic, biologic, physical processes, and water quality aspects of instream flow study. The TOD is a fairly detailed document (74 pages) with more than 2,000 pages of supplemental, highly detailed, technical appendices. The appendices contain information mostly about the water quality programs in Texas, although other topics are also covered. Appendices, background, and introductory material aside, the TOD has 8 major sections that correspond to the flowchart in the Programmatic Work Plan (PWP; see Figure 4-1): (1) study design; technical evaluations for (2) hydrology and hydraulics, (3) biology, (4) physical processes, and (5) water quality; (6) integration and interpretation; (7) study report production; and (8) monitoring and validation.

Introduction and Ecological Setting

The TOD opens with two sections, the Introduction and Ecological Setting, that present important background material that introduces the motivation of the Texas instream flow program and necessary components of an instream flow study. In the context of the state mandate to maintain a "sound ecological environment," the ecological setting of rivers is described. Biology, hydrology and hydraulics, geomorphology, water quality, and connectivity are defined and introduced as the components of an instream flow study.

Study Design

Study Design (Section 3) is a short section that identifies the major steps necessary to begin an instream flow study. Basic steps for starting an instream flow study include compiling and evaluating existing information; identifying stakeholders; identifying appropriate study areas; conducting field reconnaissance, or initial technical assessments; preliminary biological and physical surveys; and the development of geographically-specific objectives and study plans. Without much detail, this section lays out the general approach to design an instream flow study in Texas.

Hydrology and Hydraulics

By far, the Hydrology and Hydraulics section (Section 4) is the most detailed section of the TOD. In it, technical aspects of hydrologic evaluation are discussed, such as historical, naturalized, and environmental flows and flow duration curves. Examples of types of hydrologic models are mentioned. Aspects of hydraulic modeling relevant to instream flow study are major segments, too, including some guidance on how to select a representative reach and methods for data collection. One- and multiple-dimensional modeling options are detailed. Large woody debris is considered a special challenge in hydraulics, and is discussed separately in this section.

Biology

The Biology section of the TOD (Section 5) outlines with specific methods for conducting baseline surveys and understanding instream habitat. Methods for surveys of instream habitat, fish, riparian systems, and macroinvertebrates are discussed in moderate detail. The TOD Biology section describes how to sample assemblages and measure habitat conditions, calculate habitat suitability criteria, integrate calculations with simulations of aquatic physical habitat, and integrate these calculations with simulated patterns of physical habitat dynamics. Instream and riparian habitat heterogeneity are also discussed in this section.

Physical Processes

Physical processes in the TOD (Section 6) refer to hydrogeomorphic riverine processes. Compared to the previous sections, physical processes is notably brief. This section of the TOD presents compact discussions of river classification, assessment of the current status of a river in terms of its geomorphology, and sediment transport processes. Flushing flows and valley, riparian, and channel maintenance physical processes are explained. For this section, the TOD focuses primarily on describing these processes, and only scantly mentions some general methods that can be employed to assess and measure physical processes in an instream flow study.

Water Quality

Water quality is unlike the other technical aspects of instream flow study in Texas because it is regulated at the federal and state levels. There are several well established water quality programs in Texas. The TOD section on water quality (Section 7) describes the state programs and provides relevant background and administrative history of these programs. The section on water quality for instream flow studies (Section 7.3) notes that applying water quality models used in the total maximum daily load (TMDL) and Texas Pollutant Discharge Elimination System (TPDES) programs to the instream flow studies will provide consistency among state programs. Water quality models for instream flow studies, according to

Section 7.3, should take into account spatial and temporal scales; geomorphic and hydraulic conditions of the water body; and the constituents of concern. Sampling or modeling methods for instream flow studies are not presented in the TOD section on water quality.

Integration and Interpretation

Findings from the technical evaluations (i.e., biology, physical processes, hydrology and hydraulics, and water quality) will be integrated to develop a flow recommendation. The integration section of the TOD (Section 8) describes the integration process in a framework (Section 8.1; see Figure 5-1) described simply as "the steps needed to develop flow regimes." It also specifies that a quantitative analysis will be performed to identify critical relationships among the various technical aspects of an instream flow study. Instream habitat is defined as the integration of biology and hydraulics (Section 8.3) and will be predicted using a geographical information system (GIS)-based physical habitat model. The TOD presents ways in which such a model can be used. Habitat time series and habitat duration curves are described as tools for determining flow recommendations (Austin and Wentzel, 2001). The TOD stresses that many combinations of these spatial and temporal analyses can be used to identify target flow regimes. This section very briefly mentions how hydrology, physical processes, and water quality also need to be integrated into a flow regime recommendation. Quantitative Analysis (Section 8.7) includes a combination of statistical, time series, and optimization analyses. The TOD acknowledges that the "precise formulation of the instream flow optimization exercise has yet to be defined or tested," but presents examples and scenarios in which such analyses could be useful in an instream flow study. Finally, this section of the TOD briefly discusses implementation issues (Section 8.8), but does not mention how flow recommendations will be implemented administratively, scientifically, or in combination with existing Texas water statutes and regulations.

Study Report and Monitoring and Validation

The TOD ends with a very short section that states a study report (Section 9) will be produced and submitted for peer review and the final section, Monitoring and Validation (Section 10), that mentions the importance of monitoring the effectiveness of the implemented flow regime(s). The

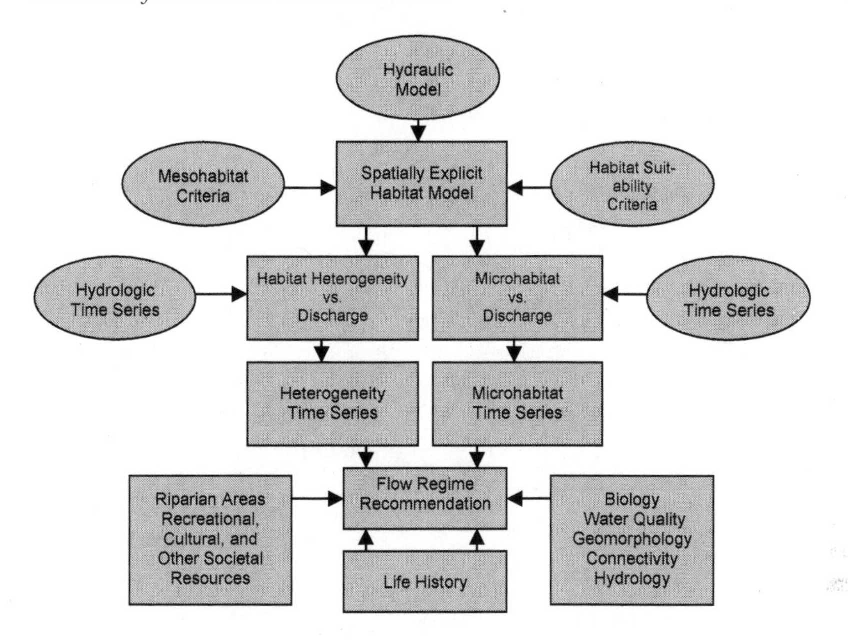

FIGURE 5-1 TOD Integration of instream flow study elements.
SOURCE: Adapted from the TOD (TPWD, TCEQ, and TWDB, 2003).

Monitoring and Validation section refers to Texas Commission on Environmental Quality's (TCEQ) surface water quality monitoring procedures and lists elements deemed important for a comprehensive monitoring program.

STRENGTHS OF AND OPPORTUNITIES TO IMPROVE THE TOD

The TOD sets out to prescribe the technical aspects, including methodologies, for conducting the detailed technical evaluations in the Texas instream flow program. This is a difficult charge to meet in a single document that is intended for diverse river subbasins across a large state. The TOD is evaluated in the following sections. The overall strengths of the document are listed first, some overarching opportunities to improve the TOD are presented next, and, finally, opportunities to improve the individual technical sections of hydrology and hydraulics; physical processes; biology; water quality; and integration and interpretation are discussed.

Strengths of the TOD

The Texas agencies are commended for drafting a document that has several strengths. The main strength of the TOD is that it encompasses the primary elements of separate technical evaluations relevant to a larger instream flow study. Technical areas of hydrology and hydraulics, physical processes, biology, water quality and connectivity are recognized as important elements and described in the TOD. The TOD also includes initial approaches for integrating results into a flow recommendation. It cannot be overstated how complicated inter-disciplinary instream flow studies can be, and Texas has made a commendable effort in designing its instream flow program to be comprehensive. The biology and hydrology and hydraulics sections reflect a commanding understanding of the relevant issues for instream flow work in Texas rivers. Finally, the TOD represents cooperation among three state agencies with separate missions. Presentations in the TOD reveal the relative expertise of each agency and hint at the promise of these three agencies working together successfully to design and implement a benchmark instream flow program in Texas.

Overarching Opportunities to Improve the TOD

The TOD is composed of several individual pieces that comprise the technical aspects of the Texas instream flow program. The TOD presents each technical piece and the process by which the pieces will be integrated (Figure 5-1) into a flow recommendation. The sharp focus of the TOD is on the distinct, technical pieces of the instream flow study; however the real challenge in instream flow science, and the weakness of the TOD, is the connections among these pieces. Landscape ecology metrics and connectivity can help strengthen these connections.

Chapter 3 outlines seven principles of a state-of-the-art program. The top three principles are to (1) preserve whole functioning ecosystems; (2) mimic, to the extent possible, a natural flow regime; and (3) expand the spatial scope of instream flow studies beyond the river channel to include the riparian corridor and floodplain systems. The whole ecosystems, natural flow regime, and the expanded spatial scale can be viewed as landscape ecology metrics of instream flow science. Used as the focus of the technical evaluations, these metrics can guide the development of instream flow recommendations. Methods by which landscape ecology elements guide an instream flow study are also listed in Chapter 3 as the last four principles: conducting studies using an interdisciplinary approach; using a variety of tools and approaches tailored to the subbasin characteristics; using adaptive management; and involving stakeholders in the process. Together, these seven principles, viewed as landscape metrics and methods, can be used to set instream flow requirements in a state-of-the-art instream flow program.

Connectivity is defined in the TOD as the "movement and exchange of water, nutrients, sediments, organic matter, and organisms within the riverine ecosystem" (TPWD, TCEQ, and TWDB, 2003). It is discussed in two paragraphs in Section 2, Ecological Setting, but not in the subsequent technical evaluation sections of the TOD. Whereas the TOD defines the concept of connectivity well, it never addresses how connectivity is a part of the Texas instream flow program. The brief, early section on connectivity lays out a nice structure in which technical evaluations could be designed or integrated; unfortunately, connectivity is not revisited in subsequent sections of the TOD. Connectivity reflects important aspects of instream flow science and the TOD should address how connectivity will be used in the detailed technical evaluations.

The dimensions of connectivity occur laterally, longitudinally, vertically, and temporally. These dimensions could be used as organizing axes for developing the conceptual models (see Table 3-2) and designing technical evaluations. Connectivity dimensions also can be used to calibrate the spatial scale of the technical evaluations to ensure compatibility and smooth integration of results. For example, the lateral dimension across a stream channel and associated floodplains could establish the spatial scale for technical evaluations of sediment erosion and deposition, flooding frequency and magnitude, aquatic and riparian species, and variation in water quality. Reconnaissance data could be collected on these assays and entered into a matrix (i.e., Table 3-2) to create a conceptual model that informs the detailed technical evaluations.

Three other overarching findings come from evaluating the TOD. First, for each technical evaluation (i.e., hydrology/hydraulics, physical processes, biology, and water quality), the TOD makes little distinction among individual basins and presents one or a very few approaches that may not be appropriate in all basins and subbasins. Of course, the TOD cannot possibly list all approaches for all possible scenarios in Texas rivers, but it should identify that a range of models, approaches, and tools may be necessary to address the highly variable characteristics of each study subbasin.

Second, considerable inconsistency is found in the level of detail among the technical sections. Some sections have highly detailed methodological processes (hydrology and hydraulics, biologic sampling), but other sections outline only very general sampling methods or none at all (integration, physical processes, and long term monitoring and validation).

Finally, many of the methods presented in the TOD lack context because measurable instream flow goals are not clearly articulated. It is understandable that the goals for the individual basin studies will vary from basin to basin and all goals cannot be identified in the TOD. Still, the TOD very briefly mentions goals (i.e., biological diversity and biological integrity) and does not discuss the methods in the context of a state-wide program or goals. Without a clearly defined goal statement or process to identify it in the PWP or TOD, the context for these technical studies is unclear.

Hydrology and Hydraulics

Overall, the section on hydrology and hydraulics demonstrates a solid understanding of the state-of-the-art in hydrologic and hydraulic methods used in the scientific and engineering community. Compared to other chapters in the TOD, Section 4 (Hydrology and Hydraulics) is quite specific about what tasks will be performed and the tools that will be used. The Texas agencies are commended for the high level of sophistication and detail presented in the hydrologic/hydraulic section of the TOD.

The weakness with the hydrologic/hydraulic TOD material is that the detailed and sophisticated methods presuppose that all techniques are applicable in all Texas basins. The hydrologic/hydraulic TOD section describes specific approaches that may not be necessary in or appropriate for all instream flow studies. A better approach to hydrology and hydraulics is to outline specific methods that could be applied in different circumstances to assure a consistent approach across the state that has enough flexibility to accommodate the variety of river systems within Texas. This improved approach will strengthen the study design phase and reduce the cost of hydrologic/hydraulic sampling and modeling by eliminating unnecessary analyses. Furthermore, connections are not explicit between the hydrologic/hydraulic techniques presented and a sound ecological environment, instream flow study goals, or the other technical disciplines of instream flow study, such as biology or water quality.

The purpose of hydraulic modeling is to define the streamflow characteristics (e.g., depths and velocities) as a function of discharge. As presented in the TOD, results from hydraulic modeling subsequently will be used to assess biological, water quality and physical processes in instream flow systems. The problem in the TOD is that these models and the results from these models are related loosely, if at all, to the other technical elements and studies. It is unclear in the TOD whether the spatial scale of the hydrologic/hydraulic studies coincides with the spatial scales of the biology, physical processes, or water quality empirical studies. A stronger connection among "the master variable," hydrology, and the other instream flow technical elements will be very important to ensure that the sophistication of the hydrologic/hydraulic tools, models, and methods is appropriate and efficient for achieving instream flow study goals. Therefore, the TOD should be revised to include explicit connections to the other technical studies to ensure that hydrologic/hydraulic technical assessments are relevant to achieving instream flow study goals, including a sound ecological environment.

The TOD and the PWP mention how hydrology is affected by human uses in the watershed. Several of the supporting documents indicate the profound effect of reservoirs on Texas rivers. Reservoirs are important considerations in Texas, as all major rivers in Texas are dammed for hydropower, municipal, or irrigation purposes. One important element missing from the hydrologic/hydraulic TOD section is a method to relate reservoir operations to instream flows. For some distance below a dam, a river's hydrology, water quality, substrate, and biota will be greatly affected by the dam's operation. The TOD also does not discuss how instream flow characteristics may change due to watershed and land use changes, such as increases in urbanization, irrigation, and impervious surface area in the watershed. Water managers have many options to affect instream flows, including dam operation, as well as issuance of water permits to withdraw water from or discharge water to a river. The TOD needs revision to consider approaches for predicting instream flow levels that take into account reservoir operations, permitting, and other watershed land uses.

Some of the two- and three-dimensional models presented in the TOD are highly sophisticated. These models require high quality input data to produce high quality, sensitive and very detailed output data about streamflow characteristics. The hydrologic/hydraulic models presented in the TOD appear too detailed for and therefore misaligned with, some of the other technical studies. For example, aquatic habitat in Texas is classified in the Aquatic Life Use scale as "Exceptional," "High," "Intermediate," or "Limited." The TOD suggests that results from hydraulic analyses can be used to broadly classify aquatic habitat in these categories. If aquatic habitat is classified in such qualitative terms, then highly quantitative outputs of hydraulic modeling may not be needed for such classification.

Another example of misalignment is with spatial scale. The accuracy of two- and three-dimensional hydraulic models is dependent on the spatial density of the data. These time- and resource-intensive models require suitably accurate input data. Some biological or geomorphic empirical studies will take place over larger spatial areas, such as over the floodplain of a segment or the home range of a key fish species. In these cases, the limited spatial scale of a hydraulic model is too fine to be of use to the geomorphic or biologic assessments. The methodologies presented in hydrology/hydraulics section of the TOD need better alignment with the other technical aspects of the instream flow studies in terms of model sophistication, sensitivity of model output, and spatial scale.

The authors of TOD Section 4 clearly have a good understanding of hydraulics, methods for gathering hydraulic data, and one- and two-dimensional flow modeling. The remaining challenge is the development of quantitative relationships between hydraulics and specific elements of a sound ecological environment. Given that rather short duration streamflow phenomena can be critical for many aquatic and riparian biota, the TOD appropriately proposes to develop naturalized flow series using daily time steps by disaggregating naturalized monthly flows from the water availability models (WAM) used for water rights permitting in Texas. Nevertheless, the TOD needs revision to make stronger connections among the naturalized flow series and biologic or other aspects of instream flow.

Summary: Hydrology and Hydraulics

Hydrology is often referred to as the "master variable" in an instream flow context because all the other aspects relate to it. Biology, physical processes, and water quality aspects of instream flow work all can be tied to components of the hydrologic regime. Indeed, the TOD and PWP suggest strongly that the intention of the Texas program is to capitalize on these naturally occurring connections to develop a strong, comprehensive instream flow program for the state. Quantifying streamflow characteristics requires highly technical methods and models, and the hydrology/hydraulics section of the TOD reflects an impressive knowledge of such approaches. Despite its level of detail and sophistication, the hydrology/hydraulics section of the TOD needs significant revision to:

- Include explicit connections to the other technical studies to ensure that hydrologic/hydraulic technical assessments are relevant to achieving instream flow study goals, including a sound ecological environment
- Consider approaches for predicting instream flow levels that take into account reservoir operations, permitting, and other watershed land uses
- Align more closely with the other technical aspects of the instream flow studies in terms of model sophistication, sensitivity of model output, and spatial scale
- Make stronger connections among the naturalized flow series and biologic or other aspects of instream flow

Biology

The general strengths of the Texas TOD section on biology include a strong introductory section that provides an excellent overview of the literature and major issues associated with choice of biological response variables and methods of data collection and analysis for instream flow recommendations. The TOD also provides an outstanding general discussion of the important issues of habitat scale, ecological processes, and species life histories. However, the TOD Biology section gives a limited description of the program's rationale and plans for implementing alternative methods for field sampling, data analysis, and derivation of flow recommendations. The connection between the biological surveys and goals of the instream flow program or of individual studies is not discussed in this section. Program elements related to biology are discussed here in the order that they appear in the TOD.

Baseline Information

The TOD baseline information section correctly identifies the starting point for a biology survey that is part of an instream flow study. The TOD highlights steps to take at the beginning of a biology sampling effort: compiling existing information, soliciting stakeholder involvement, and inventorying the types of information that will likely be needed in the biological survey (i.e., life history traits, environmental requirements, species distribution, community composition, and connectivity considerations). The TOD sets forth four types of surveys, or field reconnaissance, to be done in the process of gathering baseline information: instream habitat, fish, macroinvertebrate, and riparian surveys. All except the riparian survey section are described at a decent level of detail; the riparian survey section is brief and lists only the types of information to be considered or collected.

The field reconnaissance measures, as outlined, appear logical and sufficient, and the need for gathering and evaluating baseline information is well defended in the TOD. However, the TOD does not adequately illustrate the availability of specific data sources, the manner in which data will be gathered and analyzed, and how these analyses will influence the design and implementation of specific studies.

The TOD states that "ecological integrity" will be assessed at the reach scale, but the specific metrics for estimating ecological integrity are not identified, with the exception of using the TCEQ standard protocol for determining the appropriate Aquatic Life Use designations of surface waters. The metric developed by Texas Natural Resources Conservation

Commission (TNRCC; now TCEQ) is for statewide application and has limited ability to be tailored for specific conditions of different subbasins.

Indices of biotic integrity (IBIs; Linam et al., 2002) hold the promise of providing fast, cheap, yet comprehensive ecological indicators for long-term monitoring in instream flow studies. Given that Texas is a large state with diverse geography, climate and cultures, IBIs and other aggregate metrics must be customized for different biotic regions, river basins, and, in some cases, river segments. Regionalized IBIs that are applicable in streams of different sizes and biogeographic sub-regions would be immensely useful in the Texas instream flow program. TCEQ has recently adopted regionalized IBIs in its state-wide metric system (D. Mosier, TCEQ, personal communication, 2004) and developed these IBIs to assess water quality goals in wadeable streams. As yet, the regionalized IBIs have not been tested as a means of evaluating modified flow regimes, particularly in large, nonwadeable rivers.

Despite the strengths and promise of regionalized IBIs, they are not appropriate in all settings. Regionalized IBIs require further research and revisions to adapt the metric to apply to a range of stream sizes and orders within each region. Before being used to monitor the effectiveness of the Texas instream flow program, the Texas regionalized IBIs should be evaluated for application to instream flow studies and larger rivers. These evaluations should be published in the open, peer-reviewed scientific literature as a means to validate the Texas approach.

The choice of biotic response indicators and assessment methods should be as standardized (repeatable) as possible at all relevant spatial scales. Explicit data quality assurance/quality control and precise sampling methodologies, taxonomic identification, and quantitative methods are needed such that separate, independent groups of researchers could repeat sampling with comparable results across the state. The Biology section of the TOD should be revised to clarify biotic response indicators and assessment methods of the sampling protocol to be reliable, precise, and related to program objectives.

Instream Habitat Surveys

The TOD refers to the mesohabitat (pools, riffles, runs, rapids, and chutes) as the spatial scale for most of the biological studies and surveys. The method for designating mesohabitats is essentially visual (Vadas and Orth, 1998). Unfortunately, visual criteria and mesohabitat designations may be subjective and applied differently in different river basins. Given the TOD emphasis on mesohabitats and habitat guilds (Leonard and Orth,

1988; Vadas and Orth, 2001), this apparent subjectivity could influence the success of the Texas instream flow program. Objective criteria need to be developed to designate mesohabitats in Texas' diverse river systems. This fundamental issue needs to be addressed more thoroughly in the TOD.

Fish Surveys

The TOD discusses fish sampling with specificity, citing seines and electrofishing as primary empirical methods. These methods can be effective in some situations, but the TOD accurately mentions that they have limitations, too. For example, the TOD states that microhabitat utilization data will be collected quantitatively, but seining and electrofishing are unlikely to provide information at this fine spatial scale in any systems except the smallest streams. The TOD does not outline how these specific methods can be standardized across different size streams and various geographic regions across the state. The TOD places too much emphasis on correlational habitat suitability criteria (HSC) approaches involving fishes to the exclusion of other viable approaches and biotic components, and focuses on fish almost exclusively as aquatic fauna.

Some ecological categorizations of fish species cited by the TOD (see Linam et al., 2002; Linam and Kleinsasser, 1998) require further study and revision (e.g., the spiny-cheek sleeper is not an "omnivore," the Mexican tetra is an "omnivore", etc.). The TOD states that surveys of mesohabitats are to be conducted when flows are at or below median conditions, but the document leaves unclear how organism–habitat associations will be determined for high flow events. The problem is that a model that projects fish habitat use at median and low-flow conditions likely will not predict habitat use under high flow conditions accurately. Conversely, ecological interactions that would not occur during low-flow conditions (when there is greater habitat segregation) may occur during high flow events. Fish habitat use should be considered under base flow, subsistence flow, high flow pulse, and overbank flow conditions.

Macroinvertebrate Surveys

The TOD describes three methods for macroinvertebrate surveys in detail: kick nets, woody debris (snag), and hand-picked sampling methods. The rationale for the selection of these methods is not presented. It is implied in the text that these three methods are appropriate for all Texas rivers, when, given the range of river conditions across the states, these methods alone may not be suitable for all Texas rivers.

Riparian Area Surveys

The TOD provides limited information about selection of biological variables and survey and analysis methods for riparian habitats. The focus on connectivity of off-channel (floodplain) aquatic habitats is appropriate for some, but not all of Texas' river systems. In comparison to the fish, macroinvertebrate, and instream habitat surveys sections, the riparian survey section and presented methods are very brief. Riparian ecosystems are important components of the river ecosystem and the TOD needs to balance the treatment of riparian surveys with those of aquatic fauna. The TOD should be revised to more strongly emphasize riparian habitats as elements of a sound ecological environment, augment the methods presented for riparian surveys, and present ways to relate riparian sampling results to flow needs necessary to maintain a sound ecological environment in Texas rivers.

Instream Habitat Models

The TOD presents instream habitat models (Section 5.3) in two parts, the Quantity and Quality of Microhabitat and Habitat Heterogeneity, both of which model aquatic habitat availability in response to discharge. The TOD section on Quantity and Quality of Microhabitat (Section 5.3.1) details four steps for quantifying and qualifying these habitats: (1) sample assemblages and measure habitat conditions; (2) calculate habitat suitability criteria; (3) integrate criteria with simulations of instream habitat over a range of flows; and (4) develop habitat time series. TOD Section 5.3.2 outlines the model used to determine habitat heterogeneity.

Section 5.3 presents material that is inconsistent with other sections of the TOD. For example, macroinvertebrate sampling presented in this section is different from methods presented in the macroinvertebrate surveys section (Section 5.2.3). Section 5.3 also seems to confuse meso- and microhabitat spatial scales, sampling methods for each spatial scale, and ways to use data collected at each scale. The TOD needs careful revision to clarify inconsistencies in instream flow habitat models.

Proposed methods for calculation of habitat suitability criteria for indicator species and mesohabitat guilds seem to follow currently accepted approaches. The TOD mentions multivariate criteria for combining multiple variables (e.g., depth, velocity) simultaneously, but these criteria are not discussed. The use of habitat guilds is logical for many of the stream and river systems in Texas. However, it is likely that criteria for identification of habitat guild members will vary from basin to basin, and perhaps even from stream reach to stream reach. The degree to which habitat guild designations will be transferable between spatial units of study is unknown, and the team should consider statistical methods to estimate transferability. A reference system for habitat guilds would be useful to define mesohabitats based on biological criteria, and the derived units could be used for examining species associations in a different study system. The TOD could be revised to include such a reference system or some other means to designate mesohabitat based on biological criteria.

Modeling approaches, variables, and survey methods should be limited to the most relevant parameters consistent with available time and resources. In some cases, the potential severity of ecological risks or the complexity of the ecological setting may demand multiple approaches, some perhaps involving considerable investment of time and resources, to provide sound recommendations. Rather than attempt to apply a diverse set of methodologies to project responses by a diverse set of biological response variables at a diverse set of spatial and temporal scales, the instream flow program should develop consistent study plans using the fewest possible biological response indicators to derive defensible flow recommendations.

Summary: Biology

The Biology Section of the TOD (Section 5) gives a solid overview of the main biological considerations in an instream flow study. The strengths of the section include a strong general discussion of the important issues of habitat scale, ecological processes, and species life histories. Opportunities for improvement include strengthening connections between the detailed sampling methods and study goals; increasing consistency within the document with respect to spatial scale and sampling methods; and sharpening approaches for conducting biological surveys in dissimilar river systems across Texas. The Biology section of the TOD provides highly detailed accounts of how to conduct some sampling or modeling methods, but gives scant attention to how modeled and empirical data will be communicated, related to program goals, or integrated with other aspects of an instream flow study to derive a flow recommendation. Recommendations for addressing biological issues in the TOD include the following:

- Texas regionalized IBIs should be evaluated for application to instream flow studies and larger rivers; these evaluations should be published in the open, peer-reviewed scientific literature as a means to validate the Texas approach.
- The Biology section of the TOD should be revised to clarify biotic response indicators and assessment methods of the sampling protocol to be reliable, precise, and related to program objectives.
- Objective criteria need to be developed to designate mesohabitats in Texas' diverse river systems.
- Fish habitat use should be explored under base flow, subsistence flow, high flow pulse, and overbank flow conditions.
- The TOD should be revised to more strongly emphasize riparian habitats as elements of a sound ecological environment, augment the methods presented for riparian surveys, and present ways to relate riparian sampling results to flow needs necessary to maintain a sound ecological environment in Texas rivers.
- The TOD needs careful revision to clarify inconsistencies in instream flow habitat models.
- The degree to which habitat guild designations will be transferable between spatial units of study is unknown, and the Texas instream flow team should consider statistical methods to estimate transferability.
- The instream flow program should develop consistent study plans using the fewest possible biological response indicators to derive defensible flow recommendations.

Physical Processes

The physical processes section of the TOD (Section 6) presents riverine physical processes in four main sections: the introduction, classifying a river segment, assessing current conditions in the river, and sediment transport. Strengths of the physical processes section include: recognition of the natural and anthropogenic variability in river system status and processes, presentation of reasonable techniques for specific components of the physical processes evaluations; and acknowledgment for the need of a variety of techniques. Nevertheless, the physical processes section needs significant revision and expansion. After the hydrology component, the channel geometry may be the most important component of an instream flow study, but the TOD gives physical processes very cursory treatment. This section is noticeably shorter and less comprehensive than the hydrology/hydraulics, biology, and water quality technical segments, and it needs to be augmented to address the important issues associated with physical processes.

One major omission from the TOD is the mention of flood-dominated river regimes in Texas. Texas has a hydrological regime with high frequency of flash floods (Beard 1975). There is a spatial gradient in flash flood potential, high in west Texas and decreasing toward east Texas. Rivers with high flash flood potential may rarely or never achieve equilibrium, since channel morphology, physical habitat and flood features may be substantially rearranged in each flash flood. The geographic and geomorphic variations in flood variability are important considerations in the development of an instream flow program for the state. Different types of hydrological, hydraulic, biological, and physical assessments may be needed for river systems in different portions of the state, and criteria and expectations for instream flow management will need to accommodate dramatic difference in river hydrology across Texas. This is potentially a key factor in physical processes in Texas rivers, but it is never mentioned in the TOD.

Since physical processes vary spatially within a river system, understanding of physical processes is best constructed in a geographic context. The initial assessments should determine whether the channel and its watershed are in a state of dynamic equilibrium, and the TOD should describe processes to determine equilibrium, such as sediment budgets, models, aerial photographs and GIS, as applicable. GIS is a tool that allows efficient storage and viewing of environmental data, helps identify linkages, and improves stakeholder access to and understanding of scientific data collected in the instream flow study. While GIS is mentioned in the TOD, a general framework for GIS and its analytical role is not described. Compilation of data in a geospatial data base structure and use of a GIS to store, display, and analyze data for all parts of the study will improve the quality of the study and also provide documentation for subsequent review, reassessment and adaptive management. A GIS database should be used for data storage and analysis in instream flow studies and the instream flow program at the state level.

Physical processes vary through time. Population growth and land use change are ongoing processes that affect river systems. Changes in climate patterns over the next fifty years may also have significant effects on river discharge patterns and therefore on physical processes (U. S. National Assessment Synthesis Team, 2001). These concepts are not mentioned in the TOD. Instream flow recommendations should take into account trends in watershed and river conditions, probable future human demands on the river system, and probable future climatic change.

Classifying a River Segment

Geomorphic classification of river segments and reaches is an important component of the study that will be useful for documenting and analyzing physical processes, for selecting representative reaches and study reaches for instream habitat analysis, and for water quality analyses. Geomorphic classification provides a spatially explicit framework for analyzing physical processes and instream habitat, and a framework for selecting representative reaches for detailed analysis and modeling. Channel morphology includes a number of distinct components, including cross-section form and size, planform, slope and bed morphology (Knighton, 1998), which need to be represented in a geomorphic classification. A variety of geomorphic river classification methods have been developed since the 1980s and have been ably reviewed by Thorne (1997), Montgomery and Buffington (1998), and Kondolf et al. (2003).

The geomorphology of flood-dominated rivers presents enormous challenges for geomorphological classification, assessing the dynamic status of a river based on morphological indicators, and "maintenance flows" for sediment transport. Flood-dominated rivers can radically restructure themselves physically during individual hydrological events. This restructuring can lead to major changes in river form that serve as the basis for classification and dynamic assessment. Moreover, a flood-dominated flow regime can overwhelm attempts to maintain specific substrate and physical habitats through "maintenance" flows for sediment transport. Flash floods have real implications for changing the spatial and temporal structure and connectivity of physical habitat, both instream and in the riparian zone. Physiographic and hydrologic setting also relate to river classification. Whether reaches are gaining (water supplied by groundwater sources) or losing (supply water to groundwater sources) water can be critical in determining whether sufficient flows are provided for physical processes as well as aquatic and riparian biological needs. An oversight of the TOD is its silence on these aspects of classifying a river.

The TOD briefly describes and promotes the Rosgen (1996) method to classify streams in Texas. While the Rosgen system is widely used by land management agencies, there is considerable disagreement as to its efficacy (for example, Federal Interagency Stream Restoration Working Group, 1998; Juracek and Fitzpatrick, 2003; Kondolf et al., 2003; Miller and Ritter, 1996). A channel classification system that is hierarchical (in the sense of Bisson and Montgomery, 1996; Frissell et al., 1986) and physically-based (see Kondolf et al., 2003) may be more appropriate for instream flow studies. In addition to classification systems focusing solely on the river channel, geomorphic classification of floodplains (Nanson and Croke, 1992) and riparian zone classification (NRC, 2002c) may be useful for understanding floodplain equilibrium, channel-floodplain connectivity, and linkages between physical processes and ecological conditions.

Assessing the Current Status of the River

Assessing the status of a river reach or segment can be done qualitatively or quantitatively. Qualitative assessment can be based on morphological indicators (see Table 5-1). Morphological indicators appropriate for the study area can be developed and tested with field observations. The initial equilibrium status assessment can be verified by examining historical maps and aerial photos, or with quantitative methods, such as using channel evolution models, calculating bed level changes from gaging station records, analyzing channel width changes from historical maps and aerial photography, and assessing deposition and erosional features (McDowell, 2001; Phillips, 2003; Simon and Castro, 2003; Smelser and Schmidt, 1998). Assessing equilibrium conditions must also take into account the fundamental mode of river dynamics associated with the prevailing flood regime.

The TOD proposes identifying "any recent changes (perhaps within the last 30 to 50 years) that may have occurred to the watershed or channel" as a means to assess the current status of a river reach or segment. Trends in

TABLE 5-1 Morphological Indicators of Geomorphic Equilibrium Status

Equilibrium Status	Morphological Indicators
Aggrading	Abundant bars Low bank height to floodplain surface Active sediment deposition on floodplain surface Recently developed side channels May be braided
Incising	Very low width:depth ratio High bank height Unstable banks, failure through mass movement Bed is erosional on fine sediment, gravel, or bedrock Relatively dry floodplain with low water table
Degrading through widening	Large width:depth ratio Wide bed with little inundation in low flow season, but few active bars Armored or embedded gravel bed
Dynamic equilibrium	Intermediate in characteristics listed above

watershed land use and river conditions are relevant elements to understanding a river's dynamics, but this guidance is too general to guide consistent, repeatable technical evaluations.

Furthermore, the TOD needs to recognize that such trends can be deceptive in flood-dominated systems where individual events can drastically restructure a river system in addition to changes from on-going trends. The guidance in this section should be made more specific, and identification of the river's geomorphic equilibrium status should be included as part of the geomorphic classification.

Sediment Transport

The basic form of a river channel is a direct result of interactions among eight variables: discharge, sediment supply, sediment size, channel width, depth, velocity, slope, and roughness of channel materials (Heede, 1992; Leopold, 1994; Leopold et al., 1964). In an undisturbed watershed, there exists a dynamic equilibrium between sediment loading and the stream's capacity for sediment transport.

Sediment transport and related hydrogeomorphic processes are discussed in TOD Section 6.4, with some emphasis on valley maintenance, riparian maintenance, channel maintenance and flushing flows. The TOD appropriately underscores the importance of sediment transport and deposition among the physical processes necessary for maintaining a sound ecological environment. The TOD provides a good general discussion of sediment transport processes in Section 6.4, but fails to state how sediment transport analysis might be used in the physical process evaluations or how it relates to a sound ecological environment.

Methods for establishing a general context for sediment and its potential influences on habitat are not described. As part of the physical process evaluations, it may be necessary to quantify the sources, sinks and throughflow of sediment of different sizes—in other words, to do a reconnaissance, semi-quantitative or qualitative sediment budget (see Campbell and Church, 2003; Reid and Trustum, 2002). As land use changes within the contributing watershed, discharge levels needed for channel maintenance flows and flushing flows may also change. In such cases, a sediment budget may be needed to define current levels of sediment flux and predict future levels. Depending on the goals of the study, it may be important to develop sediment budget estimates for appropriate representative time periods, such as historic, pre-dam, agricultural, and post-dam conditions.

The TOD refers to several models that make quantitative predictions about flow and sediment transport (e.g., HEC-6, SED-2D), but model predictions about how much sediment will be yielded are subject to large errors (Simon and Senturk, 1992). This potential for error should be acknowledged in the TOD to help justify making adjustments to flushing or channel maintenance flow recommendations, if deemed necessary by monitoring and validation efforts.

The TOD recognizes three key discharge levels linked to physical processes that should be evaluated in the physical processes evaluations: floodplain maintenance, flushing flows, and channel maintenance. These terms are broad, perhaps too broad to be useful, and a more detailed breakdown of ecological and management objectives may be more helpful in instream flow studies (Kondolf and Wilcock, 1996). The four-part flow regime (i.e., subsistence flows, base flows, high pulse flows, and overbank flows) is recommended as a structure to link ecological and management objectives, physical processes, and discharge (see Table 3-2). The specific ecological or management objective that a key discharge level is intended to satisfy must be specified prior to analyzing flow requirements. The specific objectives listed in Section 6.4.4 are to restore/enhance riffle habitat; remove superficial fine sediment deposits; and remove interstitial fine sediment from gravel (Kondolf and Wilcock, 1996), but no criteria are presented in the TOD to measure progress towards achieving these objectives. In addition, the biological objectives of flushing flows should be clarified in the TOD. These biological objectives are not discussed in Section 5 (Biology) as was stated in Section 6.4.4.

For determining channel maintenance flows, sediment transport alone may not be adequate. The TOD correctly notes that flow duration, not simply instantaneous peak flow, is important in defining the channel maintenance flow (IFC, 2002). The IFC (2002) describes an empirical approach using suspended sediment rating curves, bedload rating curves, and daily discharge records to compute the channel-maintaining effective discharge (Knighton, 1998). As the TOD points out, establishing suspended and bed load rating curves requires field measurement of sediment transport at a wide range of flows, a labor intensive process that typically takes several years to complete.

An alternative approach to determine maintenance flows suggested in the TOD is to assume that bankfull stage is the minimum channel maintenance flow and occurs once every 1.5 years. This assumption is not safe. There has been a great debate on whether the 1.5-yr flow, bankfull discharge, effective discharge, and channel forming discharge are equal (Knighton, 1998). The 1.5-yr flow is bankfull for many streams, but bankfull discharge frequency can vary from less than 1 year to several tens of years, depending on the river system. Some additional alternative procedures for defining channel maintenance flows are suggested by Kondolf and Wilcock (1996).

Summary: Physical Processes

Understanding physical processes for instream flow involves consideration of hydrologic regime; channel morphology; processes that form floodplains, channels and physical habitat; sediment transport; historical alteration of the channel and floodplain; and future changes in the watershed. The TOD physical processes section identifies important problems related to geomorphology and some techniques for addressing those problems, but it does not consider the hydrologic regime in geomorphic assessments. Several important elements involved in conducting physical process evaluations are not discussed, such as the import and relevance of hydrologic regime, generally, and flood-dominated systems, specifically; GIS applications; sediment budget considerations; and impacts of changes in land use and population in the watershed over time. The strong spatial gradient in flash-flood potential and, by extension, physical processes, make necessary a wide range of assessments and tools for physical processes. This section needs significant augmentation to address the physical processes along this gradient. Currently, the TOD sets forth a thin, single set of analytical approaches for physical processes in Texas rivers that are unlikely to address the range or complexity of physical processes that exist.

Recommendations for addressing physical process concerns include:

- Augmenting this section to equal in detail the hydrology and hydraulics and biology sections and to discuss Texas hydrologic regimes, GIS applications, sediment budget methods, and impacts of changes in land use, population, and climate in the watershed over time.
- Basing instream flow recommendations on prevailing flood regimes, as well as trends in watershed and river conditions, probable future human demands on the river system, and probable future climatic change.
- Including an assessment of geomorphic equilibrium status and study of historical alterations of the channel and floodplain of the river area under study.

Water Quality

Water quality in the Texas instream flow program is treated differently than the other technical sections of hydrology and hydraulics, biology, and physical processes. Unlike the other technical aspects of instream flow, water quality is regulated by the Clean Water Act at the federal level and a number of well-established water quality programs at the state level. As specified in the statement of task, this report reviews aspects of the instream flow program relevant to the TMDL water quality program and its associated water quality models. Therefore, what follows are two sections: the evaluation of the TOD and an evaluation of the TMDL program and its associated water quality models, specifically QUAL-TX.

Evaluation of the TOD Section on Water Quality

The TOD section on water quality for instream flow studies (Section 7) notes that applying water quality models used in TMDL and Texas Pollutant Discharge Elimination System (TPDES) to the instream flow studies will provide consistency among state programs. While using the same models for multiple state programs would indeed provide consistency among them, this approach will work best if the models fulfill the needs of the instream flow goals as well as those of the water quality programs. Current water quality models can be used for discrete aspects of instream flow studies, but no current model exists that can model all water quality elements needed in an instream flow evaluation.

The TOD section on water quality contains a summary of each of the TCEQ water quality programs, and also of the Texas water availability modeling program. Summaries are presented of the Water Quality Standard and Assessment, Surface Water Quality Standards, Texas Water Quality Inventory, TMDL, TPDES, and the Water Rights Permitting and Availability programs. The primary water quality model that Texas uses, QUAL-TX, is also described. Further detail on the water quality programs is contained in three appendices to the TOD that collectively contain 26 documents and nearly 2,000 pages of material. The strength of this material is that it describes the very comprehensive structure for water quality management that the state has progressively built up over many years. A significant limitation of the TOD water quality section is that it presents a mass of documents without delineating how (1) those documents relate to the instream flow program, (2) the water quality component of an instream flow assessment should be conducted, or (3) instream flow and water quality considerations can be integrated with each other.

The appendix material clearly describes Texas' existing administrative ways to combine water quality with biology; it is done in through the Aquatic Life Use standards. All of the classified rivers and streams in Texas have a designated level for Aquatic Life Use defined in the Texas Administrative Code. Designations of Aquatic Life Use are "Exceptional," "High," "Intermediate," and "Limited." Aquatic Life Use designations rely on measurable criteria that establish levels of ecological integrity in classified water segments, including IBIs and some aspects of water quality. As coarse as these classifications are, Texas currently uses them as a systematic method of measuring aspects of a sound ecological environment in streams and rivers. Table 5-2 helps to define the attributes of these aquatic life classifications.

However, there are limitations to using Aquatic Life Use standards in instream flow studies. First, it is unclear whether the IBIs used to determine Aquatic Life Use standards are sensitive to flow variation. That is, if the flow regime changed, it is unclear whether the IBI would respond appreciably. An instream flow program may be better served by more simply defined ecological indicators that are directly related to the flow regime. Second, there are some aspects of a sound river environment that are not covered by the Aquatic Life Use component of water quality management system, such as riparian vegetation and water quality in oxbow lakes.

Finally, the Texas water quality standards account for Aquatic Life Use levels and provide a method for assessing whether these levels are being attained, but these standards do not relate aquatic life to instream flow. The Texas Water Quality Standards empower the TCEQ to protect aquatic life from degradation by pollutants but not from degradation by lack of stream flow.

If "a sound ecological environment" for instream flows differs from "ecological integrity" used in water quality assessment, the TCEQ Commissioners could be faced with two separate sets of requirements for assessing the ecological conditions of Texas streams and rivers. At a minimum, the existing Aquatic Life Use goals should be considered in implementing instream flow recommendations to avoid conflict or even establish support between the water quality and instream flow programs. Integrating the instream flow and water quality programs will provide clearer direction for all parties involved. Streamlining related programs will also reduce the potential for inconsistent or conflicting recommendations among the programs, reduce costs, and eliminate redundant analyses. Therefore, the instream flow program should be integrated with water quality, water permitting and other water-related programs in Texas.

TABLE 5-2 Aquatic Life Attributes for Aquatic Life Categories

Aquatic Life Use	Habitat Characteristics	Species Assemblage	Sensitive Species	Diversity	Species Richness	Trophic Structure
Exceptional	Outstanding natural variability	Exceptional or unusual	Abundant	Exceptionally high	Exceptionally high	Balanced
High	Highly diverse	Usual association of regionally expected species	Present	High	High	Balanced to slightly imbalanced
Intermediate	Moderately diverse	Some expected species	Very low in abundance	Moderate	Moderate	Moderately imbalanced
Limited	Uniform	Most regionally expected species absent	Absent	Low	Low	Severely imbalanced

SOURCE: TNRCC, 2000.

The Texas TMDL Program

The TMDL program in Texas[1] is the primary mechanism to remedy impairments to water quality. A TMDL is "the total amount of a pollutant a water body can assimilate and still meet state water quality standards" (TNRCC, 1999). TMDL development has come to prominence in recent years because many water bodies are not "swimmable and fishable," despite significant water quality improvements due to controls on end-of-pipe wastewater discharges. TMDLs include point- and non-point sources, such as pollution from watershed runoff, atmospheric deposition, and contaminated sediments.

The Texas TMDL program relies on water quality models that estimate nutrient, bacterial, and other pollutants in surface waters. QUAL-TX and seven other models are currently used in TMDL studies in Texas. The applicability of these models in an instream flow context, however, is untested.

Of the models used in the TMDL program (QUAL-TX; Mass Balance or CSTR; HSPF; WASP; QUAL2E; SWAT; EPIC; and EFDC), QUAL-TX has been relied upon most heavily and is therefore the focus of this discussion. QUAL-TX is a modification to the federal QUAL2E[2] model. It has been tailored for Texas river conditions, such as a site specific equation for stream reaeration. QUAL-TX is a steady state model for which the discharge is set at a small value, such as the 7 day 2 year low flow. It is most often used to estimate effects of wastewater discharge on dissolved oxygen (DO) during very low flow conditions. In some river basins, such as the San Antonio Basin, the TCEQ has developed and maintained a suite of QUAL-TX models for segments of the San Antonio River and its principal tributaries to help assess wastewater discharge permit applications.

QUAL-TX is a mainstay of the Texas wastewater discharge permitting process, and it has also been applied in about one third of the TMDL studies undertaken to date by the agency (Table 5-3). However, this model has several limitations when considering instream flow specification. Principal among them is that the model is a static or steady state model, which means

---

[1] For further information on the Texas TMDL program see the TNRCC website at *http://www.tnrcc.state.tx.us/water/quality/tmdl/index.html.*

[2] The QUAL model was originally developed in Texas and later further developed and adopted for national use by EPA.

TABLE 5-3 Use of Water Quality Models in TMDL Studies in Texas

Model	Number of TMDL studies
QUAL-TX	9
Mass Balance or CSTR	4
HSPF	3
WASP	3
QUAL2E	2
SWAT	2
EPIC	1
EFDC	1
No model used	5
Total	30

SOURCE: Data from G. Rothe, TCEQ, personal communication, 2004.

it operates only for a single streamflow discharge, but an instream flow assessment has to consider a whole range of flows that may occur and their time patterns of occurrence. QUAL-TX operates on river or stream segments ¼ mile to 1 mile long. The model yields an average dissolved oxygen value for a river reach that contains many mesohabitat zones and associated aquatic communities. QUAL-TX also accounts for spatial variations in water quality between water in the center of a stream and that along the banks, and the vertical variations in dissolved oxygen content with depth.

Another limitation of QUAL-TX is that it assumes a flat-bed stream, i.e., the bottom area of the stream does not change as the flow approaches zero. A real stream has spatially varying bed topography; therefore higher areas of the stream bed are exposed and become dry as flow diminishes, and lower parts remain submerged longer than would be if the bed were flat. Although some of the other water quality models are dynamic and offer a greater range of possible bed geometries than QUAL-TX, the other models used in the TMDL program still employ spatial computational units of the same order of size as QUAL-TX. It might be possible through field scale research to quantify the spatial and temporal variations of dissolved oxygen within a reach so that with a daily and reach-averaged dissolved oxygen concentration available, some type of "down scaling" process could be applied to infer the spatial and temporal patterns of dissolved oxygen at the meso- and microhabitat scale.

In the four part instream flow regime (subsistence flows, base flow, high pulse flows, and overbank flows) (see Figure 3-2, Table 3-2), QUAL-TX may be useful in establishing the subsistence flow. In other words, QUAL-TX could estimate the flow needed to maintain minimum water quality standards. Since depressed DO impairs Aquatic Life Use in many streams in Texas, QUAL-TX may be a useful means of examining what flows are needed to maintain adequate levels of dissolve oxygen during low flow conditions.

However, the water quality component of an instream flow technical evaluation should involve other aspects, as well, including suspended sediment, temperature, and other water borne nutrients and pollutants. The flow regime and the various constituents of water quality act together to produce a sound ecological environment and these aspects need to be considered when defining instream flow requirements for a particular river.

Ideally, a water quality and temperature simulation model for instream flow assessment needs would allow for:

- Time varying hydrology across the full range of flow variation from floods to drought low flows
- The effect of management variations such as alternative strategies for releasing water from reservoirs
- watershed processes for sediment production and nonpoint source pollution generation
- point sources of pollution from wastewater discharges
- instream processes of chemical transformation and sediment transport
- local scale variations in flow and water quality characteristics within stream mesohabitats and microhabitats

There is no single simulation model currently available which can perform all of these functions. A mechanism is needed to combine hydrologic, water quality and hydrodynamic models across spatial scales to achieve this range of capabilities. The emerging technology of Hydrologic Information Systems is providing some capabilities that could contribute to this goal (Maidment, 2002).

Summary: Water Quality

While just as important as hydrology and hydraulics, biology, and physical processes in a Texas instream flow program, water quality is treated differently than its sibling components. Water quality is a regulated entity in Texas and has a well established set of state and federal programs. The TOD ably describes these programs. These programs, administered by TCEQ, meet their purposes of ensuring that surface waters in Texas comply with regulatory standards. Instream flow considerations are not the focus of the state's water quality programs. Therefore, the instream flow program's elements that contend with water quality must be aligned with the existing water quality programs, so as to avoid conflicting requirements for maintaining sound ecological environments in Texas rivers. The TOD presents more than 2,000 pages of water quality material. A significant limitation of the TOD water quality section is that it does not refer to this material or discuss how the water quality component of an instream flow assessment should be conducted, or how instream flow and water quality considerations can be integrated with each other.

The Texas TMDL program's aim is to improve water quality in Texas surface waters. In total, eight water quality models are used in the TMDL program, with QUAL-TX used more than the seven others. QUAL-TX is a steady state model that models DO. QUAL-TX is applicable to instream flow studies in that (1) DO is an important constituent of water quality that strongly influences aquatic biology; (2) QUAL-TX has an established record of use in Texas, including use with Aquatic Life Use designations; and (3) it operates on the same spatial scale as many biological sampling efforts. The primary limitations of the model include (1) it models DO for only one discharge at a time and instream flow studies require water quality measurements over a range of flows; and (2) DO is only one constituent of water quality, when others, such as suspended sediment, also influence a sound ecological environment. The eight models used in the TMDL program can address discrete pieces of water quality as it relates to instream flow studies, but none can simulate all of the aspects that need to be included in a comprehensive instream flow technical evaluation.

The water quality component of the Texas instream flow program reflects the existing strengths of the Texas water quality management program. These strengths include a comprehensive water quality database for Texas streams and rivers, an established set of water quality standards, and procedures for assessing compliance with them, including standards for aquatic life. The TOD does not define how these existing procedures will need to be adapted or refined for use in an instream flow assessment. A significant limitation of the bioassessment component of the existing water quality program is the lack of a comprehensive database of empirical biological information compared to extent and history of the data maintained by the TCEQ on water quality.

The major findings and recommendations for the water quality component of the TOD are:

- The TOD, with appendices, presents thorough documentation of the Texas water quality programs, but does not outline how this program can be integrated with or used in an instream flow program.

- QUAL-TX is a steady state model that can accurately model DO for a single rate of flow, a limitation for a comprehensive instream flow technical evaluation. However, there is no single simulation model currently available which can model all instream flow functions, and a mechanism is needed to combine hydrologic, water quality and hydrodynamic models across spatial scales.
- A more comprehensive method is needed for storing all the biological and physical data acquired during Aquatic Life Use assessments, and a more complete digital inventory of biological data on the past condition of Texas streams and rivers needs to be compiled.
- The instream flow program should be integrated with water quality, water permitting and other water-related programs in Texas to avoid conflict or establish support between the water quality and instream flow programs.

Integration and Interpretation

Integration of the results of the hydrology, biology, water quality and physical process investigations into flow recommendations is critical to the success of any instream flow study. This is a very difficult task because the methods for integration are not well documented (see Chapter 3), and too often, the individual investigations are not designed to be integrated with each other. The Integration and Interpretation section of the TOD (Section 8) presents a process (Figure 5-1) to derive a flow recommendation that uses instream habitat models to integrate hydraulics and biology. The Integration section of the TOD has an in-depth review of quantitative analyses and a very brief section about hydrology, water quality, and physical processes integration.

The TOD describes a vague integration process that is based on several assumptions. It is assumed that relevant aspects of aquatic habitat can be modeled by habitat models. It is also assumed that a standard set of techniques and models will be applied in all river basins. The TOD states that integration is to be accomplished mainly through quantitative analyses, but these analyses are not described in enough detail or in context to guide consistent, repeatable studies in river basins across the state. These assumptions need to be better explained and defended in the TOD to provide much needed support for the integration process presented.

The main weaknesses of the integration section are that (1) it is represented by a complicated Integration Framework (Figure 5-1) that is never thoroughly explained; (2) it does not mention the goals of the study as part of the integration process; (3) the process of how technical evaluations (Sections 4, 5, 6, and 7) are used to derive flow recommendation is not well described; and (4) the integration of biology and hydraulics is given far more attention than the other technical aspects of water quality, physical processes, and hydrology. While the purpose of integration is to pull all of the elements together, this section of the TOD ultimately stands alone. With few exceptions, the material in this section makes no reference to the technical evaluations of the previous sections of the TOD, and the previous TOD sections do not mention that results from sampling efforts will ultimately be used in the Integration Framework.

That said, the integration phase of any instream flow study is decidedly the most difficult, and methods to integrate several interdisciplinary studies into a single flow recommendation are not well documented in the current literature (IFC, 2002; Postel and Richter, 2003). Examples of possible approaches to Integration are the Building Block or Percent-of-Flow (Flannery et al., 2002) approach (see Chapter 3). The Building Block approach essentially builds a recommended instream flow hydrograph, or set of hydrographs, using key pieces of information developed during technical studies. The percent-of-flow approach uses results from the technical evaluation to determine appropriate levels of allowable flow depletion (typically expressed as percentages of the natural flow) during different times of the year, or during different water year types. These are just two approaches to integrating the various aspects of instream flow studies. Other approaches are being used and developed, but very few are well documented.

Integration should be conceived early in the study design phase to ensure that studies fit together conceptually and are aligned with each other and program and subbasin goals. This type of integration between disciplines may require different models than those models routinely used within disciplines, different sampling methodologies, or sampling at different spatial scales. In these and maybe other ways, the state agencies may have to adopt new approaches to data collection and analyses, since the agencies normal intra-disciplinary practices may not lead to an integrated approach.

The Framework

Arguably, the Integration Framework (Figure 5-1) which is intended to illustrate "the steps needed to develop flow regimes" is the most critical element of the TOD's integration process. However, the framework in the TOD is presented very briefly in one short paragraph; the framework figure does not indicate any order of sequence; and the boxes of the framework contain general topics, not "steps," towards integrating across several disciplines and technical evaluations. The Integration Framework is very difficult to navigate. Furthermore, the framework figure omits program or study goals, without which, the purpose of and connections among the integration efforts and goals is obscure. In order to be useful, the Integration Framework must be described more thoroughly in the text and/ or revised to articulate the specific steps to be taken or specific points of consideration in the process of developing an instream flow recommendation.

Instream Habitat

The Instream Habitat sub-section (Section 8.2) describes how GIS-based physical habitat models, hydraulic models, and habitat time series can be used to integrate hydraulics and biology for instream flow purposes. The TOD presents these models as a menu of options that can be used separately or in combination to identify flow regimes in all river basins, but it does not give guidance as to under which circumstances each model is most appropriate. The models are adequately described in terms of what function each model fulfills; however their position(s) in the flowchart and methods used to derive a flow recommendation are not explained. This section states that instream habitat models will be used with output from the hydraulic and biological technical evaluations, but the earlier technical sections of the TOD do not indicate that their results or output will be compatible with these instream habitat models. The instream habitat models focus exclusively on integrating hydraulics with biology, and leaves unclear whether any models can be used to integrate hydrology, water quality, and physical processes.

Quantitative Analysis

Aside from a passing reference to statistical and time series analyses, the section on quantitative analyses (Section 8.7) focuses almost exclusively on optimization analyses. Optimization analysis is proposed as a technique to "identify and evaluate alternative flow conditions that maximize, or at least preserve, ecological health" (TPWD, TCEQ, and TWDB, 2003). The goal of optimization is to make a "best" or optimal decision. Optimization has the benefits of being quantitative and leading to a single alternative; however optimization has significant shortcomings as a primary method for reaching an instream flow recommendation. A main shortcoming is that optimization is a mathematical function that cannot easily include broad ecological, legislative or social goals in its syntax. Furthermore, the type of optimization presented in the TOD "has yet to be defined or tested."

Hydrology, Biology, Physical Processes, and Water Quality Sections

These four sections, plus a fifth, Other Integration Considerations, are very brief recaps of the important elements that should be included in an integration exercise. These sections are too brief to be useful in guiding integration processes and need significant augmentation.

Summary: Integration and Interpretation

With little doubt, the integration phase is the most difficult and least documented phase in instream flow science. Most often, the purpose of this phase is to pull together results from different technical evaluations into a single flow recommendation, but the process could be more efficient if integration is conceived in the early study design phase and focuses on common goals or objectives. Various models (GIS-based physical habitat models, hydraulic models, habitat duration curves) and quantitative tools can be helpful to derive a flow recommendation, and some of those tools are introduced in this section. The Integration and Interpretation section of the TOD needs significant revision to:

- correspond more strongly to the methods presented in the biology, hydrology and hydraulics, water quality and physical processes sections of the TOD;
- revise the Integration Framework to include sequential steps and clearer direction of how to combine results from the technical evaluations with appropriate models to derive flow recommendations; and
- augment sections on integrating Hydrology (Section 8.3), Water Quality (Section 8.4), Physical Processes (Section 8.5) and Other Integration Considerations (Section 8.6) to equal in detail and application those presented in Instream Habitat (Section 8.2).

SUMMARY AND RECOMMENDATIONS

The TOD sets out methods for the technical evaluations of hydrology and hydraulics, biology, physical processes, and water quality in the Texas instream flow program. The Texas TOD (1) makes little distinction among individual basins and sets forth a standardized set of tools for use in river basins that are highly variable across the state; (2) is inconsistent in the level of detail among the four technical sections; (3) encompasses the primary elements of separate evaluations relevant to a larger, instream flow study, those of hydrology and hydraulics, biology, physical processes and water quality, with tenuous connections among them and vague associations to an instream flow recommendation; and (4) presents methods that lack context because measurable instream flow goals are not clearly articulated.

Therefore, the TOD is recommended to be revised to:

1) strengthen linkages among individual studies on instream biology, hydrology and hydraulics, physical processes, and water quality, and stronger connections between studies and components of flow regime;

2) include greater capacity for and reference to site-specificity at the (sub) basin-scale;

3) design the biological, physical processes water quality, and hydrology and hydraulics instream flow studies at commensurate spatial and temporal scales to improve the ability to integrate findings from the various technical evaluations into a single flow recommendation;

4) strengthen the physical processes section to align more closely with the hydrology and hydraulics and biology sections;

5) clarify methods and the flowchart in the Integration and Interpretation section;

6) describe how connectivity will be used in the Texas instream flow studies;

7) augment the monitoring and validation (i.e., adaptive management) section to monitor progress towards meeting the stated goals; and

8) establish means to set goals for the individual studies that relate to the state-wide definition of a sound ecological environment.

In general, the major findings and recommendations for each technical section are as follows:

1) Hydrologic and hydraulic technical studies reflect a significant understanding of hydrology, hydrologic measurements, and analyses commonly required for performing instream flow studies. The TOD presents highly sophisticated yet standardized hydrologic and hydraulic analyses. Not all models, however, will fit all streams and the analytical approaches should be more closely tailored to the specific objectives of the instream flow study.

2) The physical processes section is notably brief, especially in comparison to the hydrology and hydraulics and biology sections. It omits discussions about Texas hydrologic regimes as they relate to physical processes, GIS applications, sediment budget estimates, and impacts of changes in land use, population, and climate in the watershed over time. The physical processes section needs to be expanded to be comparable to the hydrology and biology sections and include discussions on Texas hydrologic regime, GIS application, sediment budget analyses, and impacts of land use, populations and climate changes in the watershed.

3) Texas regionalized IBIs should be evaluated for application to instream flow studies and larger rivers; these evaluations should be published in the open, peer-reviewed scientific literature as a means to validate the Texas approach.

4) The instream flow program should be integrated with water quality, water permitting and other water-related programs in Texas to avoid conflict between the water quality and instream flow programs.

5) The Integration Framework (TOD Figure 8.1) needs be revised to include sequential steps and clearer description of the proposed process to derive flow recommendations from combining results from the technical evaluations with appropriate models.

6

Implementation Issues

Implementation may be the most important step in any instream flow effort. It is included in the original Programmatic Work Plan (PWP) framework (Figure 4-1) and mentioned in the Technical Overview Documents (TOD), but the Texas instream flow documents hardly address the critical issue of how the instream flow recommendations will be implemented. Implementation issues will be especially important to the Texas instream flow program because it is expected that the state and its citizens will take a number of years to develop and refine mechanisms for instream flows and sound ecological environments in the state's highly diverse river systems.

Federal and state environmental policies counsel proactive efforts by states to protect instream values. The federal and state Clean Water Acts set broad and ambitious goals for the protection of fishable and swimmable waters nationally. Effective effluent limitations and ambient water quality standards established under these laws depend upon certain minimum base flows. Similarly, the federal Endangered Species Act can significantly constrain water resources management when species found in a waterway are listed as threatened or endangered. Yet, experiences in places like the upper Colorado River basin in Colorado and Utah and on the Platte River in Nebraska suggest that it is possible to conserve fish and wildlife by protecting instream flow regimes and taking other conservation measures while allowing for water resource development. Pro-active conservation efforts that prevent an endangered species listing are almost always less onerous and less resource intensive than is the work needed to conserve and recover a species once it is listed. The same can be said for river health—it is generally easier to protect or maintain a river's status than to restore a degraded river to a previous or improved condition.

Implementation will occur at two levels in the Texas state-wide instream flow program. First, the state-wide program will be implemented as the river basin studies are conducted and completed. Second, instream flow recommendations developed for specific river systems must be implemented. There will be challenges in implementing both the program and the recommendations for flow regimes in specific river systems.

This chapter outlines some considerations for instream flow implementation. In regard to implementation of specific flow regimes in specific river systems, the chapter (1) discusses approaches and challenges related to balancing human and ecosystem needs, (2) provides some examples of instream flow work, and (3) briefly discusses the use of models in implementation. This chapter also highlights the importance of adaptive management and on-going peer review, and considers some of the technical recommendations from previous sections of the report in the context of implementing both the state-wide program and flow recommendations for specific river systems.

BALANCING HUMAN AND ECOSYSTEM NEEDS

A major aspect of implementing an instream flow recommendation requires a deft balance in allocating water among disparate and competing uses. This balance between human and ecosystem needs is reflected in the PWP statement of finding a flow regime that conserves fish and wildlife and human uses of water. Allocating water for a range of water needs and uses is a challenge in many places across the United States. In Texas, specifically, situations exist that further upset this delicate balance, such as the state's groundwater withdrawal policies and rapidly changing land uses, the state's many reservoirs, over-allocated rivers compared to rivers where water remains available, non-priority river basins, and climate changes.

Anticipating Changes in Groundwater Withdrawal and Land Use in the Watershed

Groundwater

Groundwater is a critical aspect of instream flow. Springs and seeps contribute a significant portion of the total water that flows in many of the state's rivers and streams and illustrate how groundwater and surface water function as a unified hydrologic system in many instances. Well pumping can influence groundwater discharge to rivers and streams, with the potential to alter subsistence and base flow conditions. Even though significant, unregulated withdrawals from underground water sources could affect instream flows in significant ways, Texas's system for the allocation of surface and groundwater resources is legally distinct. This disjunction between the unified physical nature of surface and groundwater systems and the bifurcated allocation system for surface- and ground-waters raised some questions about the efficacy of instream flow recommendations that may be affected by groundwater withdrawals.

Hopefully, the new and more aggressive framework for managing underground water resources that was established by Senate Bill 1 will prove effective in integrating these two interrelated water resources as Senate Bill 1's new framework is woven into the state's larger water resources allocation system. Nevertheless, efforts to establish instream flows on surface rivers and streams could be significantly and adversely affected by future groundwater withdrawals without better integration in protecting these two resources. At a minimum, groundwater models and other tools can be used to assess influences of groundwater pumping on surface flows.

Land Use

Changes in land use also can have a marked effect on a watershed's hydrologic behavior, and thus may need to be when protecting or restoring instream flows. As a watershed is converted from its natural vegetative cover into urban areas or farms, infiltration capacity of the watershed is reduced, leading to increases in high flow pulses and overbank flows and decreases of subsistence and base flows. A sound instream flow recommendation will need to anticipate these types of future changes in hydrologic conditions, so that water managers can implement necessary modifications to water management practices or make permitting decisions consistent with instream flow goals.

Rivers with Large Reservoirs

Opportunities exist for achieving instream flow goals, especially in basins with large storage or hydropower dams. Where permitting activities allow, dam operations can be used to release targeted instream flows (see Savannah River example, Box 6-2). Implementing instream flow recommendations on rivers that are heavily influenced by dam operations will typically require cooperation among the state, stakeholders, and dam managers to integrate instream flow goals with other dam management purposes. Increasingly, water managers, river conservationists, and other stakeholders are exploring ways to modify dam operations to improve releases of water for ecological and recreational benefits in downstream river ecosystems (Postel and Richter, 2003; Richter et al., 2003). The use of dams and reservoirs for flood control, water supply, hydropower generation, or recreation in some cases can impose constraints on opportunities for hydrologic restoration, but in many cases some flexibility will exist to change dam operations to improve downstream conditions and serve the original purposes of the dam. Because large dams can have considerable influence on river flows for tens to hundreds of miles downstream, improved dam operations can benefit long stretches of river.

The large number of federally-influenced dams in Texas, and their distribution across many different river basins, suggest considerable opportunity for partnership with appropriate federal entities and other partners in attaining instream flow goals through improved dam operations. For example, the U.S. Army Corps of Engineers (USACE) is now collaborating with The Nature Conservancy to evaluate opportunities for modifying operations of USACE dams across the country to improve river health. Under the "Sustainable Rivers Project," fourteen dams on ten rivers are being assessed for flow restoration opportunities, with the expectation that this number will grow considerably in coming years. Similar partnerships with federal dam managers and river basin authorities are encouraged in Texas. By working closely with the USACE, Bureau of Reclamation, river basin authorities and other dam managers, significant progress toward instream goals can be realized in many of the state's river basins.

In Texas, there may be considerable opportunity to influence the operation of non-federal hydropower dams, particularly when these facilities are applying for re-licensing under the Federal Energy Regulatory Commission (FERC). Instream flow goals for rivers influenced by these hydropower dams can be communicated through participation in FERC relicensing processes.

Rivers Where Water Remains Available

Water in river basins with un-allocated water presents the opportunity for water to be set aside in some way to attain or maintain instream flow goals. One way to capitalize on available water is through direct appropriations. Direct appropriations have been effectively used in other states (i.e., Colorado [Colo. Rev. Stat. Ann. § 37-92-102(3), 2004]) and may offer a potential guide for states like Texas. Another way to use available water for instream flow purposes is through increased efficiency measures. Other states have encouraged water uses to implement efficiency measures, and Texas could use these existing examples as a guide for its instream flow program. For example, the state of Oregon permits water users to salvage water by implementing conservation measures, but requires that twenty-five percent of the salvaged water be made available to the stream (Or. Rev. Stat. § 537.470(3), 2003). The Texas approach to direct and indirect reuse may encourage water users to adopt conservation measures that make water available for both consumptive and non-consumptive uses. Finally, where feasible, the reservation of unappropriated water has the potential for preserving the state's flexibility while it makes decisions about competing demands for water.

Over-allocated Rivers

Sometimes, instream flow recommendations may exceed, or even significantly exceed, available flows. In these cases, innovation is required to protect instream values. Texas is not the only the only state where rivers are over-appropriated; this is the situation in many parts of the West. Thirteen other western states have put in place statutory or administrative strategies (e.g., Colo. Rev. Stat. Ann. § 37-92-102(3), 2004; Utah Code Ann. § 73-3-3(11), Supp. 2004; Wyo. Stat. Ann. § 41-3-1001-1014, 2003) for instream flow protection. These sister state programs reinforce the notion that instream flow programs can be implemented even in highly arid regions. Other states' programs also represent a significant reservoir of experience and expertise that Texas policy makers can consult in moving forward (e.g., continued participation in the Instream Flow Council). One of the lessons learned in other western states with more established programs is that a water right or other device for protecting instream flows that is junior to consumptive water rights is limited in its effect. The passage of time accentuates that problem as more and more water is appropriated. On over-appropriated rivers, delay will likely only exacerbate the policy choices facing the state to protect instream flows. On rivers that are not fully appropriated, delay may present fewer opportunities in the future, or opportunities that could be attained only at significantly greater cost than today.

The Texas Water Trust[1] within the Texas Water Bank is an entity with significant potential: it could facilitate willing buyer/willing seller transactions in which senior consumptive water rights could be acquired and converted to instream uses, either for a term of years or in perpetuity. The state and the Trust also could examine statutory measures that are being used in Colorado (Colo. Rev. Stat. Ann. § 37-80.5-104.4, 2004), Montana (Mont. Code Ann. § 85-20-1001 et seq., 2004), and other western states

---

[1] The Texas Water Trust is established by Texas Water Code Ann. § 15.7031.

where water users or conservationists enter into agreements for dry-year leases with senior water rights holders to maintain flows in a waterway.

Approaches to Instream Flows in Non-priority Basins

The Texas instream flow program has identified six priority river basins to initiate the instream flow program. These priority basins represent a small subset of the total number of rivers and streams in the state, and the state may wish to expand the instream flow program to other rivers as it develops instream flow experience. For this expansion, it may be desirable to have some sort of methodology for setting priorities. The Lyons Method or Consensus Criteria for Environmental Flow Needs are good options, but have some limitations (see description of these programs in Chapter 3 and in the following section on Model Use). Ideally, a priority-setting methodology would help water managers determine the order in which additional rivers will be evaluated for instream flow recommendations and weigh a range of alternatives to maximize the state's future opportunities to protect adequate instream flows.

Texas water documents (TWDB, 2002a) and testimony given at open meetings in Austin and San Antonio suggest that existing current water rights cannot be satisfied fully during periods of below-average flows in many river segments. Problems created by low-flow situations may be compounded by the projected increases in population and water demand in Texas. Over the next 50 years, the Texas population is projected to grow dramatically. By the year 2050, as many as 900 cities will need to either reduce demand or develop new water sources in order to meet projected needs during drought or low-flow conditions (TWDB, 2002a). A potentially important consideration in a program with tiered implementation, such as the instream flow program in Texas, is that as demand for water increases, it will be difficult to implement instream flow recommendations on second and third-tier rivers.

In the interim, before those second- and third-tier rivers can be studied, Texas may want to consider options that preserve its flexibility to be able to meet future needs on rivers that are not yet considered priority basins. Preserving the status quo, especially on important rivers, may be important at least until the initial period is over and focus can be turned to non-priority river systems' instream flow requirements. One way to preserve flexibility may be through permits, as was done in the permit for the Guadalupe River in the City of Victoria where instream flows were protected through innovative permitting (see Box 6-1).

Planning for Climate Change

Texas historically has experienced significant drought cycles that have complicated water providers' efforts to meet human needs. National and international scientists who study the phenomenon of climate change have concluded that climatic perturbations may exacerbate that existing drought cycle and consequently reduce the amount of water available for both human needs and ecosystems beyond what has been observed in the period of record. This trend is particularly worrisome for a state that anticipates dramatic population and economic growth over the same time period during which atmospheric scientists anticipate that the effects of climate change will begin to manifest themselves. The combination of population and economic growth and an intensified drought cycle may seriously stress river systems and the water supplies available for both human and instream needs. It could force adoption of mitigation measures, such as conservation and efficiency as well as even more aggressive drought planning than the state already has undertaken. In addition, reduced precipitation, increased evaporative losses, and reduced storage all would act to reduce minimum flows that are integral elements of both effluent standards and water quality standards. One potential result of climate change would be significantly increased costs to provide potable water to human populations and for agricultural production.

IMPLEMENTATION EXAMPLES

Despite the challenges of balancing human and ecosystem needs in implementing an instream flow recommendation, many examples exist of how it has been done successfully. Most commonly, an instream flow recommendation is aimed to either protect some existing instream value or restore flow to a targeted value. In protection mode, managers need to guard against changing hydrologic conditions beyond the thresholds represented by the instream flow recommendations. In restoration mode, managers need to bring back hydrologic conditions to a desired condition. Three examples are presented that show how instream flow recommendations were implemented for protection or restoration purposes. The first example shows how instream flow goals could be attained through permitting activities; the second highlights the importance of flow variability in instream flow recommendations; and the last example shows how models can be used to restore targeted flows.

The first example is from Texas. In Chapter 3, the "percent of flow" approach is described as a way to determine instream flow recommendation values as appropriate levels of allowable flow depletion (typically expressed as percentages of the natural flow). The innovative language of a permit issued by the Texas Natural Conservation Commission (TNRCC, now TCEQ) to the City of Victoria in 1996 for withdrawals from the Guadalupe River affords a significant degree of protection to instream flows (including subsistence flows, base flows, high flow pulses, and overbank flows; see Box 6-1) with the effect of protecting instream flows in a manner similar to the "percent of flow approach." Although not initially intended as an instream flow effort, the City of Victoria permit shows how instream flow recommendations could be implemented in Texas.

The Savannah River in Georgia and South Carolina provides another example relevant to Texas. The Savannah River is a managed river with large dams, not unlike many rivers in Texas. Dam operations were used to maintain in-channel flow in the river all year at higher than natural levels (Table 6-2). River systems are dynamic, and more water in the river is not always "better" for the river ecosystem. High flows in the Savannah kept floodplain soils too moist too consistently for floodplain trees to reproduce. The Savannah River example shows the importance of flow variability, not merely presence of water in the channel, in protecting riverine ecosystems.

The Upper Peace River in Florida shows how a flow restoration project worked using hydrologic models. Here, hydrologic simulation models were used to explore potential strategies for recovering instream flow conditions to a targeted level. In the Upper Peace River, a number of water and land use strategies are being employed for the purpose of recovering subsistence and base flows to targeted levels (see Box 6-2).

MODEL USE IN INSTREAM FLOW IMPLEMENTATION

Hydrologic simulation models that estimate hydrologic changes associated with future development are extremely useful in designing water management strategies and can help water managers assess the potential effectiveness of attaining instream flow goals. Different types of hydrologic simulation models can be used to assess changes in watershed runoff, groundwater flow, or reservoir operations. When applied interactively, these models can estimate cumulative interaction of these water and land use changes on a river's hydrologic regime. These interactions can be used to determine the likelihood or degree that hydrologic changes relate to instream flow requirements.

BOX 6-1
Example of a Texas Water Permit with Instream Flow Protection

In May 1993, the City of Victoria, Texas, applied to the Texas Natural Resources Conservation Commission for a permit to build a 1,000 acre-feet capacity off-channel reservoir and to divert up to 20,000 acre-feet of water per year from the Guadalupe River to fill the reservoir to be a water supply source for the city. The permit was granted by the Commission in January 1996, with restrictions as to how the water could be withdrawn from the river.

The permit defines the annual volume of water which can be diverted (20,000 acre-feet per year), and also the rate at which it can be diverted depending on the observed discharge at the USGS stream gage for the Guadalupe River at Victoria (Gage 08176500). The following restrictions apply:

- When the observed discharge is at or above the "normal" flow level, the diversion can be up to 150 cfs, where normal flow is defined in Table 6-1 below.
- When the diversion of water would reduce the flow below normal, the diversion is limited to the difference between the observed flow and normal flow plus 10% of the remaining flow, the total diversion not to exceed 150 cfs.
- When the flow at the gage is below normal, the diversion is limited to 10% of the gaged flow.
- When the observed discharge drops below the "low" flow, diversion must cease. The "low" flow is defined in Table 6-1.

TABLE 6-1 Flow Statistics for the Guadalupe River at Victoria by Month

Month	Normal Flow (cfs)	Low Flow (cfs)
January	387	150
February	440	150
March	660	200
April	687	250
May	1260	200
June	995	250
July	540	300
August	414	300
September	490	200
October	353	150
November	357	150
December	374	150

SOURCE: Adapted from data presented in City of Victoria 1996 water withdrawal permit. Provided by Steve Densmore, TCEQ, 2004.

The method used to define normal flows is a combination of the Lyons Method based on monthly median flows, and ecological flow needs of the Guadalupe Estuary as follows: "normal flows, based on gaged records, will be described as 40% of the monthly median streamflow in the months of October through February; 60 percent of the monthly median flow in months of March, April, July, August and September; and a flow rate for the months of May and June based on a prorated share of the minimum flow values calculated to maintain beneficial inflows for the living resources and ecological integrity of the Guadalupe Estuary." Moreover, the "low flow" is defined as the "amount of flow for each month needed to protect water quality in the river, and to a limited extent provide, on a short-term basis, dissolved oxygen levels for sustaining fish and wildlife species." Figure 6-1 provides a graphical interpretation of these diversion limitations.

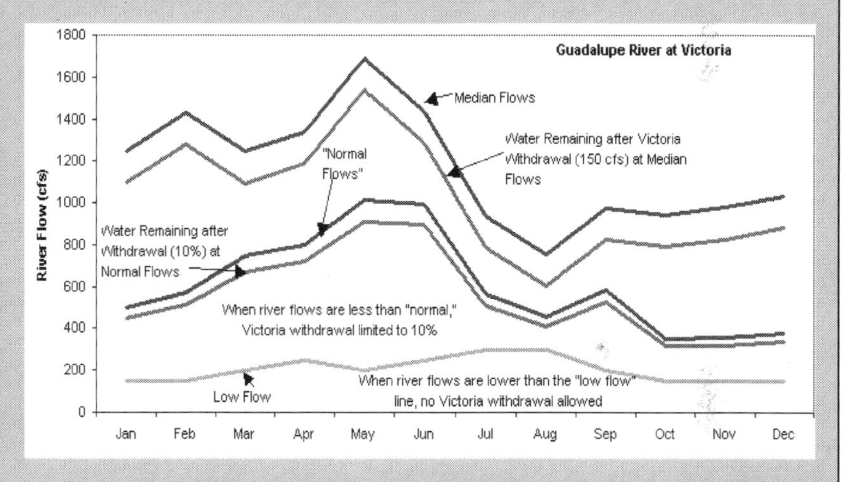

FIGURE 6-1 Graphical representation of diversion limitations for the Guadalupe River at Victoria, TX.

TABLE 6-2 Savannah River Comparison of Existing Conditions with Instream Flow Recommendations[2]

	Existing Conditions	Instream Flow Recommendations
Base Flows:		
January	5,190 – 12,320 cfs	7,500 – 12,000 cfs
February	5,200 – 13,350 cfs	7,500 – 13,500 cfs
March	5,500 – 12,500 cfs	7,500 – 13,500 cfs
April	5,850 – 13,000 cfs	7,500 – 13,500 cfs
May	5,790 – 13,100 cfs	6,200 – 13,500 cfs
June	7,040 – 13,330 cfs	6,200 – 8,500 cfs
July	5,700 – 13,000 cfs	6,200 – 8,500 cfs
August	4,950 – 13,050 cfs	5,500 – 8,500 cfs
September	4,930 – 13,200 cfs	5,500 – 8,500 cfs
October	4,700 – 12,030 cfs	5,500 – 9,000 cfs
November	4,880 – 11,540 cfs	6,200 – 9,000 cfs
December	5,210 – 10,060 cfs	6,200 – 9,000 cfs
High Flow Pulses:		
Magnitude	0 – 34,500 cfs	16,000 – 30,000 cfs
Frequency	0 – 11 events per year	2 – 6 events per year
Overbank Flows		
Magnitude	> 50,000 cfs	50,000 – 70,000 cfs
Frequency	7 in 50 years (1:7)	once every 3 years (1:3)

SOURCE: Existing condition data from USGS gaging station #02198500 near Clyo, GA.

Water availability models have been developed by the TCEQ for each of the 23 major river basins in Texas. These models are used to assess whether sufficient water remains available within each basin to satisfy existing surface water withdrawal permits and instream flow requirements as estimated with the Lyons Method (see Chapter 3). Sometimes, however, the Lyons Method can generate instream flow estimates that are less than half of the average base flows in some months of the year (see Box 6-3). The Lyons Method in water availability modeling may also result in underestimation of the instream flow needs that might be defined in a more detailed instream flow study. Another possible incompatibility exists between Lyons Method-based water availability models and the type of instream

[2] These existing and recommended flows pertain to the "floodplain reach" of the Savannah River. Instream flow recommendations for dry, average, and wet years have been lumped in this table for simplicity. It is clear from the monthly base flow summaries that current base flow conditions occasionally drop below the targeted levels, and at other times are higher than specified by the instream flow recommendations.

flow recommendations contemplated in this report. The water availability models operate on monthly time steps, but instream flow recommendations are commonly based upon daily targets or withdrawal limits, or include high flow pulse or overbank flow recommendations intended to last only a few hours to days (see Box 6-3). For purposes of statewide water planning and water permitting in basins for which detailed instream flow studies have not been conducted, a statistical hydrology method may better characterize normal monthly base flow and high flow conditions.

Therefore, the current water availability models could be reviewed to determine whether they can operate on daily time steps in addition to or in lieu of the current monthly time steps. This review could also evaluate water availability model utility in exploring a broad range of water management and restoration options, interactions of surface and groundwater systems, and if necessary, other computer tools to enable assessments of strategies for attaining instream flow goals.

ADAPTIVE MANAGEMENT

The crux of adaptive management is to learn from early instream flow studies and make changes, accordingly, as more information is amassed. In order to do so, it will be critically important to put in place a systematic and consistent mechanism for monitoring flow levels and biological responses. A set of ecological indicators responsive to streamflow variations and a systematic monitoring program for these indictors can help to adaptively manage and chart progress towards maintaining a sound ecological environment for each river. These indicators could also be monitored in rivers statewide to track changes and measure progress towards maintaining a sound ecological environment in Texas.

It is anticipated that much will be learned from the application of the recommended approach during the early years of the program. It should be expected that the Texas agencies will want to modify the final study framework or specify different kinds of initial technical assessments or detailed technical studies in subsequent instream flow studies. With state-level oversight of subbasin studies, information gleaned from earlier studies can be shared and discussed and as necessary, modified, for future activities.

BOX 6-2
Restoring Instream Flows on the Upper Peace River, Florida

The Florida Water Act of 1972 directed the state's water management districts to set "minimum flows and levels" for all streams, rivers, and natural lakes to ensure that water withdrawals do not result in significant harm to water resources and ecological health. When existing conditions or 20-year projections suggest that targeted instream flows or lake levels will be violated, the Water Act requires that a recovery or prevention strategy be developed.

When instream flow requirements for the upper Peace River were established in 2002, the Southwest Florida Water Management District (SWFWMD) realized that a recovery strategy would need to be implemented to restore necessary base flow conditions. Groundwater withdrawals from the Floridian aquifer, primarily for agriculture and phosphate mining, have lowered the potentiometric surface by 30-40 feet in the aquifer. These groundwater declines have resulted in several detrimental impacts to the water resources of the area, including the cessation of flow from in a major spring and reductions in Peace River base flows.

The recovery strategy for the upper Peace River includes a variety of measures designed to reduce existing demands or augment available supply. The measures to be implemented in the Peace River watershed have been selected after using hydrologic simulation models to evaluate the cost-benefit ratios of a large number of possible restoration options. The SWFWMD has estimated that the selected restoration projects could provide as much as 75 cfs of additional flow to the upper Peace River during a 90-day low flow period. Some of these measures include:

CONTINUING REVIEW OF THE PROGRAM

An instream flow program has scientific parts that nest within a policy context. It is particularly important for the program and for the recommendations that the scientific aspects be as free of technical dispute as possible. Close access to and open communication with a wide range of technical experts on instream flow science can help assure that the science is and remains objective and at the state of the science. A valuable role for scientists who are not directly working on studies within the instream flow program is to review the sampling methodologies, results of the individual technical studies, and progress of the overall instream flow program. Results from these reviews can be communicated to the involved scientists, instream flow scientific community at large, and stakeholders. Review by an independent group of scientists will help track the progress and efficacy

Water Conservation – Many different water conservation strategies are being implemented in the urban, agricultural, industrial, and mining sectors. These strategies include increased use of reclaimed wastewater, a variable water fee structure based on volume of use, education, and other demand management initiatives.

Flow Enhancement – More than 30,000 acres of un-reclaimed phosphate mine lands exist within the SWFWMD, much of which causes surface runoff in the watershed to pond in settling areas or pit lakes instead of contributing to Peace River flows. Some of these areas will be reconnected to the river; others will be used as reservoirs that will store runoff during periods of high flow and subsequently release water to the river during periods of low or no flow.

Wetland Restoration – 20,000 acres of wetlands that were historically altered or destroyed by agricultural activities will be restored by acquiring fee interest or conservation easements on the lands and then restoring their natural hydrologic functions. This is expected to improve surface water storage in floodplain areas during floods, enhance aquifer recharge, and improve base flow conditions.

SOURCE: SWFWMD 2002, 2004

of the instream flow program over time, just as the initial peer review was designed to provide, "the highest level of confidence for all interested and affected parties that the framework within which these studies will be carried out is scientifically sound." In order to fulfill this comprehensive program objective that involves scientists from a variety of disciplines, state agencies, and other stakeholders, the creation of an independent, interdisciplinary, periodic peer review process for the instream flow program is recommended.

POLICY CONTEXT FOR TECHNICAL RECOMMENDATIONS

In the proposed revised instream flow framework (Figure 4-2), the technical aspects of conducting an instream flow study are couched between two policy actions: setting goals and implementing the instream flow

BOX 6-3
Estimating Instream Flow Needs with Hydrologic Desktop Methods

A number of methods exist for estimating instream flow needs when little or no ecological information is available to define ecosystem water requirements. Some of these approaches—called "hydrologic desktop methods"—are based upon statistical characterizations of historic or naturalized flow data. The Lyons Method, used in Texas for surface water permitting, is an example of a hydrologic desktop method. Using the Lyons Method, monthly instream flow requirements are estimated by computing the medians of all daily flows for each month, and then multiplying those monthly medians by a specified factor. For October through February, this factor is 0.40; for other months, a factor of 0.6 is applied.

Hydrologic desktop methods can be very useful in obtaining a ballpark estimate of instream flow needs in rivers for which detailed instream flow studies have not yet been conducted. However, they must be applied carefully to ensure that they generate instream flow estimates that are consistent with instream flow protection goals. For example, in Figure 6-2, the median base flow levels for each month for the Guadalupe River at Victoria are shown, along with a line representing the average high flow pulse level.[3] The base flow values have been estimated using a "base flow separation technique" which separates the river's base and subsistence flows from high flow pulses and overbank flows caused by rainfall events.

The Lyons Method would protect much of the base flow in some months, but in other months would leave much of the base flow unprotected (Figure 6-2). The uneven levels of base flow protection afforded by the Lyons Method are in part attributable to the different factors that are applied to monthly medians as described above. Use of monthly medians in a hydrologic desktop method can also yield inconsistent degrees of protection for base flows. Monthly medians are computed using all river flows during the month – base flows, high flow pulses, and even floods are all rolled into the calculation of a monthly median. As a result, it is often hard to predict how closely the median, or a method like Lyons, will compare to base flows.

The Consensus Criteria for Environmental Flow Needs (CCEFN), adopted by the Texas agencies in 1997 in their guidelines for regional water plan development, suffers from this same shortcoming. The CCEFN provide three different levels of instream flow protection, depending upon estimates of what the naturalized monthly flow would have been.[4] The flow protection

---

[3] Based upon USGS gaging station records for 1935-2002. Base flows and high flow pulses are computed by using a base flow separation technique in the "Indicators of Hydrologic Alteration" (IHA) software that separates base flows from high flows in the daily discharge record.

[4] The CCEFN protects the monthly median level when naturalized flows are greater than or equal to the monthly median; protects the 25th percentile level when naturalized flows are less than the median but greater than the 25th percentile; and protects a fixed threshold of flow (such as 7Q2) when naturalized flows are less than the 25th percentile.

offered by the CCEFN can differ considerably from the Lyons Method because the CCEFN is based upon model-calculated estimates of "naturalized" monthly flows instead of the measured historic flows used in the Lyons Method. While the CCEFN do provide some protection for high flow pulses or floods in addition to base flows, they would still protect only about half of the average high flow pulse levels in the Guadalupe River, as shown in Figure 6-2.

In sum, hydrologic desktop methods such as the Lyons Method or CCEFN that are based on monthly medians may lead to inconsistent and unreliable protection of base flows while generally under-protecting high flow pulses and overbank flows. Hydrologic desktop estimates can be improved by first applying a base flow separation analysis to the daily data series, and then computing estimates of normal base flows or high flows separately, as illustrated in Figure 6-2. For example, if the instream flow goal is to protect base flows from excessive depletion, an instream flow target can be developed using the base flow median, or some fraction thereof. If a certain number of high flow events are to be protected as well, these can be added to the base flow estimates.

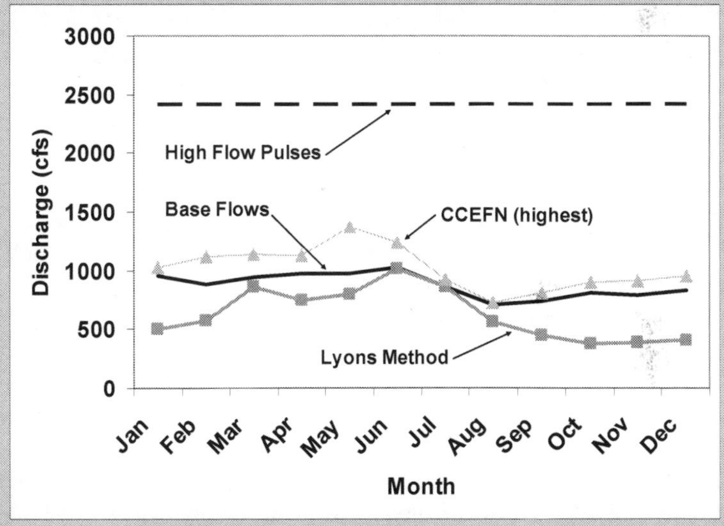

FIGURE 6-2 Comparison of base flows, Lyons Method estimations, and CCEFN[5] for the Guadalupe River at Victoria.

---

[5] CCEFN data source: Kathy Alexander, TCEQ, personal communication, 2004.

recommendation. This graphical representation illustrates a reality of instream flow programs—the science exists in a policy context. Likewise, the technical recommendations proffered in this report also exist in a policy context. This section relates some of the previous recommendations to the implementation aspects of an instream flow program and provides a policy context for some of the technical recommendations.

A major recommendation of this report is the presentation of a statewide context for individual subbasin studies with **two levels of oversight**: one at the state level for management and program consistency and one at the subbasin level for goals and approaches that are tailored to the specific needs of the study basin. While the Texas Water Development Board (TWDB), TCEQ, and the Texas Parks and Wildlife Department (TPWD), are key players in statewide water resources planning, a number of other state-created agencies involved with water resources management may have a legitimate interest in becoming involved with instream flow studies on rivers and streams within their jurisdiction (Baty, 1999).

The TWDB, TPWD, and TCEQ have developed significant expertise in instream flow science and have demonstrated a clear ability to work cooperatively on a complex and sensitive set of issues. As a result, it seems reasonable that these three agencies would remain responsible for the oversight of the instream flow program, in order to assure (1) that all studies are conducted in conformance with the final instream flow study framework and (2) that methodologies used in setting instream flow recommendations are consistent.

In Chapter 4, the recommendation is made that the role and degree of **stakeholder involvement** should be clarified. The most likely instream flow recommendation to be implemented is one where interested parties participated in the formulation of instream flow goals, provided input on study design, and were briefed on data collected or assembled during the studies. Early and frequent public participation in the instream flow process can be critical to the study's success, and consistent commitment to public participation can materially enhance the likelihood and acceptance of a flow recommendation's implementation.

From the review of the water quality models used in the Texas total maximum daily load program, the recommendation was made that the instream flow program should be **integrated** with the water quality, water permitting, and other **water-related programs** in Texas. Several water-related programs already exist at the state level, including those associated with water quality, streamflows, bays and estuaries, and water permitting. Some of these are overlapping regulatory and planning processes and all of them could have a bearing on instream flow requirements. The instream flow program can build upon or augment these programs. To the extent that the instream flow program uses methods and approaches that are consistent with existing programs, both decision makers and stakeholders will have a clear understanding of how the programs can work together to strengthen overall water management in Texas.

SUMMARY

This chapter presents some of the practical aspects, and challenges, of implementing instream flow programs and recommendations. The act of implementing an instream flow program or study requires deft balance among disparate and competing uses for river water. This balance includes considerations of groundwater, watershed and land uses, planning in an era of climate change and under a range of available (or scarce) water. Three examples of successfully implemented instream flow recommendations underscore a range of important issues such as using permitting to achieve instream flow goals; the importance of flow variability in implementing instream flow recommendations; and use of hydrologic simulation models in flow restoration projects. Hydrologic models play an important role in instream flow science (see Chapter 3) and their role in implementation is described in this chapter. Hydrologic simulation models and water availability models both have relevant uses in instream flow implementation. The limitations of some "hydrologic desktop" methods are also discussed.

Large-scale, state-wide instream flow programs, like the one in Texas, are often implemented over a number of years. In these cases, it is expected that the instream flow managers will learn from the early studies and apply those lessons to subsequent studies. Adaptive management strategies allow for modifications in methods or implementation due to more or updated information. As per adaptive management, it is anticipated that much will be learned during the early years of the instream flow program, and the Texas agencies will likely modify the final study framework, and implementation of initial technical assessments or detailed technical studies as the program and studies mature.

Over the life of the Texas instream flow program, and through adaptive management, many changes may be made to instream flow methodologies, implementation, or goals of the program. The scientific integrity of the instream flow program through these changes must not be compromised. Review by an independent group of scientists will help track the progress and efficacy of the instream flow program, methodologies, and results from individual studies over time. In order to fulfill this comprehensive program objective that involves scientists from a variety of disciplines, state agencies, and other stakeholders, the creation of an independent, interdisciplinary, periodic peer review process for the instream flow program is recommended.

Recommendations made earlier in the report are presented in the context of implementing the Texas instream flow program and instream flow recommendations. Specifically, this report recommends that the Texas program have two levels of oversight: one at the state-level for overall program consistency and one at the subbasin level for individual studies. Given the expertise and demonstrated ability to work cooperatively, this chapter observes that the TWDB, TPWD, and TCEQ are well poised to maintain the state-level of oversight for the instream flow program. Stakeholder involvement is discussed in Chapters 3 and 4 in the context of setting goals and building public support for instream flow work; stakeholder involvement in this chapter is acknowledged as important element in realizing the implementation of an instream flow recommendation. Finally, a recommendation was made in Chapter 5 to integrate the instream flow program with water quality, water permitting and other water-related programs in Texas. To the extent that the instream flow program uses methods and approaches that are consistent with existing programs, both decision makers and stakeholders will have a clear understanding of how the programs can work together to strengthen overall water management in Texas.

RECOMMENDATION

The creation of an independent, interdisciplinary, periodic peer review process for the instream flow program is recommended.

References

Austin, B. and M. Wentzel. 2001. Two-dimensional fish habitat modeling for assessing instream flow requirements. Integrated Water Resources Management 272: 393-399.

Baker, V. 1977. Stream-channel response to floods with examples from central Texas. Geological Society of America 88: 1057-1071.

Baty, C. 1999. A Commentary on Texas Water Law and Policy. Natural Resources Journal 39: 121.

Beard, L. 1975. Generalized Evaluation of Flash Flood Potential. Center for Research in Water Resources Report CRWR-124. Austin, TX: University of Texas.

Bisson, P., and D. Montgomery. 1996. Valley Segments, Stream Reaches, and Channel Units. In Methods in Stream Ecology, eds. F. Hauer and G. Lamberti. San Diego, CA: Academic Press.

Bounds, R. and B. Lyons. 1979. Existing Reservoir and Stream Management: Statewide Minimum Streamflow Recommendations. Austin, TX: Texas Parks and Wildlife Department.

Bovee, K., B. Lamb, J. Bartholow, C. Stalnaker, J. Taylor, and J. Henrikson. 1998. Stream habitat analysis using the Instream Flow Incremental Methodology. U.S. Geological Survey Information and Technical Report USGS/BRD-1998-0004. Fort Collins, CO: USGS Biological Resources Division.

Calow, P. and G. Petts. 1992. The Rivers Handbook: Hydrological and Ecological Principles. Volume 1. Boston, MA: Blackwell Scientific Publications.

Calow, P. and G. Petts. 1994. The Rivers Handbook: Hydrological and Ecological Principles. Volume 2. Boston, MA: Blackwell Scientific Publications.

Campbell, D., and M. Church. 2003. Reconnaissance sediment budgets for Lynn Valley, British Columbia: Holocene and contemporary time scales. Canadian Journal of Earth Sciences 40(5): 701-713.

Federal Interagency Stream Restoration Working Group. 1998. Stream Corridor Restoration: Principles, Processes and Practices. National

Engineering Handbook no. 653. Washington, D.C.: USDA Natural Resources Conservation Service.

Flannery, M., E. Peebles, and R. Montgomery. 2002. A percent-of-flow approach for managing reductions of freshwater inflows from unimpounded rivers to southwest Florida estuaries. Estuaries 25(68): 1318-1332.

Frissell, C., W. Liss, C. Warren, and M. Hurley. 1986. A hierarchical framework for stream habitat classification: Viewing streams in a watershed context. Environmental Management 10:199-214.

GAO (The Government Accountability Office). 2004. Environmental Indicators: Better Coordination is Needed to Develop Environmental Indicator Sets hat Inform Decisions. GAO 05-52. Washington, D.C.: GAO.

Hayes, M. 2002. Major River Basins in Texas. Austin, TX: Texas Water Development Board.

Heede, B. 1992. Stream dynamics: An overview for land managers. General Technical Report RM-72. Fort Collins, CO: U.S. Forest Service, Rocky Mountain Forest and Range Experiment Station.

IFC (Instream Flow Council). 2002. Instream Flows for Riverine Resources Stewardship: Revised Editon. United States: IFC.

Jager, H., H. Cardwell, M. Sale, M. Bevelhimer, C. Coutant, and W. Van Winkle. 1997. Modeling the linkages between flow management and salmon recruitment in streams. Ecological Modeling 103:171-191.

Jager, H., J. Chandler, K. Lepla, and W. VanWinkle. 2001. A theoretical study of river fragmentation by dams and its effects on white sturgeon populations. Environmental Biology of Fishes 60:347-361.

Juracek, K., and F. Fitzpatrick. 2003. Limitations and implications of stream classification. Journal of the American Water Resources Association 39: 659-670.

King, J. and D. Louw. 1998. Assessment of instream flow requirements for regulated rivers in South Africa using the Building Block Methodology. Aquatic Ecosystem Health and Management 1:109-124.

Knighton, D. 1998. Fluvial Forms and Processes. London: Arnold.

Kondolf, G. and Wilcock, P. 1996. The flushing flow problem: Defining and evaluating objectives. Water Resources Research 32(8): 2589-2599.

Kondolf, G., D. Montgomery, H. Piégay, and L. Schmitt. 2003. Geomorphic classification of rivers and streams. In Tools in Fluvial

Geomorphology, eds. G. Kondolf and H. Piégay. Chichester, England: John Wiley & Sons.

Kraft, M. and S. Furlong. 2004. Public Policy: Politics, Analysis, and Alternatives. Washington, D.C.: CQ Press.

Leonard, P. and D. Orth. 1988. Use of habitat guilds of fishes to determine instream flow requirements. North American Journal of Fisheries Management 8:399-409.

Leopold, L. 1994. A View of the River. Cambridge, MA: Harvard University Press.

Leopold, L., M. Wolman, and J. Miller. 1964. Fluvial Processes in Geomorphology. San Francisco, CA: W. H. Freeman.

Linam, G. and L. Kleinsasser. 1998. Classification of Texas freshwater fishes into trophic and tolerance groups. River Studies Report 14. Austin, TX: TPWD.

Linam, G., L. Kleinsasser, and K. Mayes. 2002. Regionalization of the Index of Biotic Integrity for Texas Streams. Texas Parks and Wildlife Department (TPWD) River Studies Report No. 17. Austin, TX: TPWD.

Maidment, D. 2002. Arc Hydro: GIS for Water Resources. Redlands, CA: ESRI Press.

McDowell, P. 2001. Spatial variation of channel morphology at the segment and reach scales, Middle Fork of the John Day River, northeastern Oregon. In Geomorphic Processes and Riverine Habitat, eds. J. Dorava, et al. American Geophysical Union, Water Science and Application Series, Vol. 4. Washington, D.C.: American Geophysical Union.

Meyer, J., M. Alber, W. Duncan, M. Freeman, C. Hale, R. Jackson, C. Jennings, M. Palta, E. Richardson, R. Sharitz, J. Sheldon, and R. Weyers. 2003. Summary Report Supporting the Development of Ecosystem Flow Recommendations for the Savannah River below Thurmond Dam. Online. Available *http://outreach.ecology.uga.edu/publications/pdf/summaryreport.pdf*. Accessed February 14, 2005.

Miller, J. and J. Ritter. 1996. Discussion: An examination of the Rosgen classification of natural streams. Catena 27: 295-299.

Montgomery, D. and J. Buffington. 1998. Channel processes, classification and response. In River Ecology and Management: Lessons from the Pacific Coastal Ecoregion, eds. R. Naiman and R. Bilby. New York, NY: Springer-Verlag.

Mosier, D. and R. Ray. 1992. Instream Flows for the Lower Colorado River: Reconciling Traditional Beneficial Uses with the Ecological Requirements of the Native Aquatic Community. Austin, TX: Lower Colorado River Authority.

Nanson, G. and J. Croke. 1992. A genetic classification of floodplains. Geomorphology 4: 459-86.

NRC (National Research Council). 1990. A Review of the USGS National Water Quality Assessment Pilot Program. Washington, D.C.: National Academy Press.

NRC. 1999. Downstream: Adaptive Management of Glen Canyon Dam and the Colorado River Ecosystem. Washington, D.C.: National Academy Press.

NRC. 2000. Ecological Indicators for the Nation. Washington, D.C.: National Academy Press.

NRC. 2002a. The Missouri River Ecosystem: Exploring the Prospects for Recovery. Washington, D.C.: National Academy Press.

NRC. 2002b. Opportunities to Improve the U.S. Geological Survey National Water Quality Assessment Program. Washington, D.C.: National Academy Press.

NRC. 2002c. Riparian Areas: Functions and Strategies for Management. Washington, D.C.: National Academy Press.

NRC. 2004a. Confronting the Nation's Water Problems: The Role of Research. Washington, D.C.: National Academies Press.

NRC. 2004b. Review of the U.S. Army Corps of Engineers Restructured Upper Mississippi-Illinois River Waterway Feasibility Study. Washington, D.C.: National Academies Press.

NRC. 2004c. Adaptive Management for Water Resources Project Planning. Washington, D.C.: National Academies Press.

Orsborne, J. and C. Allman, eds. 1976. Proceedings of the Symposium and Specialty Conference on Instream Flow Needs. Volumes I and II. Bethesda, MD: American Fisheries Society.

Phillips, J. 2003. Toledo Bend Reservoir and geomorphic response in the lower Sabine River. River Research and Applications 19: 137-159.

Poff, N., J. Allan, M. Bain, J. Karr, K. Prestegaard, B. Richter, R. Sparks, and J. Stromberg. 1997. The natural flow regime: A paradigm for river conservation and restoration. BioScience 47:769-784.

Postel, S. and B. Richter. 2003. Rivers for Life: Managing Water for People and Nature. Washington, D.C.: Island Press.

Railsback, S., R. Lamberson, B. Harvey, and W. Duffy. 1999. Movement rules for spatially explicit individual-based models of stream fish. Ecological Modeling 123(2-3):73-89.

Railsback, S., and B. Harvey. 2002. Analysis of habitat selection rules using an individual-based model. Ecology 83:1817-1830.

Reid, L. and N. Trustum. 2002. Facilitating sediment budget construction for land management applications. Journal of Environmental Planning and Management 45(6): 865-887.

Rhodes, K. and C. Hubbs. 1992. Recovery of Pecos River fishes from a red tide fish kill. Southwestern Naturalist 37:178-187.

Richter, B., R. Mathews, D. Harrison, and R. Wigington. 2003. Ecologically sustainable water management: Managing river flows for ecological integrity. Ecological Applications 13:206-224.

Rosgen, D., 1996. Applied River Morphology. Pagosa Springs, CO: Wildland Hydrology, Inc.

Simon, A., and J. Castro. 2003. Measurement and analysis of alluvial channel form. In G. Kondolf and H. Piégay, eds. Tools in Fluvial Geomorphology. Chichester, England: John Wiley & Sons, Ltd.

Simon, D. and F. Senturk. 1992. Sediment Transport Technology. Littleton, CO: Water Resources Publications.

Smelser, M. and J. Schmidt. 1998. An assessment methodology for determining historical changes in mountain streams. USDA Department of Agriculture Forest Service, Rocky Mountain Research Station, General Technical report RMRS-GTS-6.

SWFWMD (Southwest Florida Water Management District). 2002. Upper Peace River: An Analysis of Minimum Flows and Levels. Brooksville, FL: SWFWMD.

SWFWMD. 2004. Southern Water Use Caution Area Recovery Strategy. Brooksville, FL: SWFWMD.

Tennant, D. 1976. Instream flow regimens for fish, wildlife, recreation, and related environmental resources. Fisheries 1(4):6-10.

Tharme, R. and J. King. 1999. Development of the Building Block Methodology for Instream Flow Assessments. South Africa Water Research Commission Report No.576/1/98.

Thorne, C. 1997. Channel types and morphological classification. In Applied Fluvial Geomorphology for River Engineering and Management, eds. C. Thorne, R. Hey, and M. Newson. Chichester, Eng-land: John Wiley and Sons.

TNRCC (Texas Natural Resource Conservation Commission). 1999. Developing Total Maximum Daily Load Projects in Texas: A Guide for Lead Organizations. GI-250. Austin, TX: TNRCC.

TNRCC. 2000. Texas Surface Water Quality Standards. Chapter 307 of the Texas Administrative Code, Adopted by the Texas Natural Resource Conservation Commission, July 26, 2000.

TPWD, TCEQ, and TWDB (Texas Parks and Wildlife Department, Texas Commission on Environmental Quality, Texas Water Development Board). 2002. Texas Instream Flow Studies: Programmatic Work Plan. Online. Available *http://www.twdb.state.tx.us/InstreamFlows/pdfs/Programmatic_Work_Plan.pdf*. Accessed February 14, 2005.

TPWD, TCEQ, and TWDB (Texas Parks and Wildlife Department, Texas Commission on Environmental Quality, Texas Water Development Board). 2003. Texas Instream Flow Studies: Technical Overview, Draft. Online. Available *http://www.twdb.state.tx.us/InstreamFlows/pdfs/TechnicalOverview-Draft080803.pdf.* Accesed February 14, 2005.

TWDB. 2002a. Water for Texas—2002. Online. Available *http://www.twdb.state.tx.us/publications/reports/State_Water_Plan/2002/FinalWaterPlan2002.asp.* Accessed February 14, 2005.

TWDB. 2002b. Exhibit B: Guidelines for Regional Water Plan Development. Online. Available *http://www.twdb.state.tx.us/rwpg/twdb-docs/Data%20Guidance%20072302-modified.pdf.* Accessed February 14, 2005.

U.S. National Assessment Synthesis Team. 2001. Climatic Change Impacts on the United States: The Potential Consequences of Climate Variability and Change. Report for the U.S. Global Change Research Team. Cambridge UK: Cambridge University Press.

Vadas, R. and D. Orth. 1998. Use of physical variables to discriminate visually determined mesohabitat types in North American streams. Rivers 6:143-159.

Vadas, R. and D. Orth. 2001. Formulation of habitat suitability models for stream fish guilds: Do the standard methods work? Transactions of the American Fisheries Society 130:217-235.

Ward, J. and J. Stanford. 1983. The serial discontinuity concept of lotic ecosystems. In Dynamics of Lotic Ecosystems, eds. T. Fontaine and S. Bartell. Ann Arbor, MI: Ann Arbor Science Publishers.

Ward, J. 1989. The four-dimensional nature of lotic ecosystems. Journal of the North American Benthological Society 8(1): 2-8.

Waters, T. 1995. Sediment in streams: Sources, biological effects, and control. American Fisheries Society Monograph 7. Bethesda, MD: American Fisheries Society.

Acronyms

AFY	Acre feet per year
CCEFN	Consensus Criteria for Environmental Flow Needs
cfs	Cubic feet per second
CWA	Clean Water Act
DO	Dissolved oxygen
DTS	Detailed Technical Studies
EPA	United States Environmental Protection Agency
ESA	Endangered Species Act
FERC	Federal Energy Regulatory Commission
FIFS	Framework for Instream Flow Studies
GIS	Geographical information system
HSC	Habitat suitability criteria
IBI	Index of Biotic Integrity
IFC	Instream Flow Council
IFIM	Instream Flow Incremental Methodology
ISF	Instream flow
IHA	Indicators of Hydrologic Alteration
ITA	Initial Technical Assessment
LCRA	Lower Colorado River Authority
LWD	Large woody debris
NPDES	National Pollution Discharge Elimination System
NRC	National Research Council
PHABSIM	Physical Habitat Simulation Approach
PWP	Programmatic Work Plan
QA	Quality assurance
QC	Quality control
SWFWMD	Southwest Florida Water Management District
TCEQ	Texas Commission on Environmental Quality
TMDL	Total maximum daily load
TNRCC	Texas Natural Resources Conservation Commission
TOD	Technical Overview Documents
TPDES	Texas Pollutant Discharge Elimination System
TPWD	Texas Parks and Wildlife Department

TSS	Total suspended solids
TWDB	Texas Water Development Board
USACE	United States Army Corps of Engineers
USGS	United States Geological Survey
WAM	Water Availability Model

Appendix A

Glossary

Adaptive management—A process through which management decisions can be changed or adjusted based on additional information.

Aggradation—(1) Geomorphic process in which sediment is carried downstream and deposited in streambeds, floodplains, and other water bodies resulting in a rise in elevation in the bottom of the water body. (2) The occurrence when the supply of sediment is deposited and stored in the active channel.

Allocation—See Water allocation.

Alluvial stream—A stream with a bed and banks of unconsolidated sedimentary material subject to erosion, transportation, and deposition by the river.

Appropriation—A specified amount of water set aside by Congress, other legislative body or state or provincial water regulatory authority to be used for a specified purpose at a specified place, if available.

Aquatic life—All organisms living in or on the water.

Bankfull discharge—The discharge at channel capacity or the flow at which water just fills the channel without over-topping the banks.

Base flow—Average streamflow in the absence of significant precipitation or runoff events. Also known as "normal flow."

Bedload—Material moving on or near the streambed.

Bedload discharge—The volume of bedload passing a transect in a unit of time.

Beneficial use—A cardinal principle of the prior appropriation doctrine. It has two components: the nature or purpose of the use and the efficient or non-wasteful use of water. State constitutions, statutes, or case law may define uses of water that are beneficial. Those uses may be different in each state, and the definition of what uses are beneficial may change over time.

Bypass—(1) A channel or conduit in or near a dam that provides a route for fish to move through or around the dam without going into the turbines. (2) That stream reach below a dam that is essentially skirted by the flow used to generate electricity.

Channel—That cross section containing the stream that is distinct from the surrounding area due to breaks in the general slope of the land, lack of terrestrial vegetation, and changes in the composition of the substrate materials.

Channelization—The mechanical alteration of a natural stream by dredging, straightening, lining, or other means of accelerating the flow of water.

Connectivity—Maintenance of lateral, longitudinal, and vertical pathways for biological, hydrological, and physical processes.

Discharge—The rate of streamflow or the volume of water flowing at a location within a specified time interval. Usually expressed as cubic meters per second (cms) or cubic feet per second (cfs).

Diversion—A withdrawal from a body of water by human-made contrivance.

Drainage area—The total land area draining to any point in a stream. Also called catchment area, watershed, and basin.

Flood—Any flow that exceeds the bankfull capacity of a stream or channel and flows out on the floodplain.

Floodplain—(1) Land beyond a stream channel that forms the perimeter for the maximum probability flood. (2) A relatively flat strip of land bordering a stream that is formed by sediment deposition.

Flow—(1) The movement of a stream of water or other mobile substance from place to place. (2) Discharge.

Flow regime—The distribution of annual surface runoff from a watershed over time such as hours, days, or months (See also Hydrologic regime).

Fluvial—Pertaining to streams or produced by river action.

Gradient—The rate of change of any characteristic, expressed per unit of length. (See Slope.) May also apply to longitudinal succession of biological communities.

Groundwater—In general, all subsurface water that is distinct from surface water; specifically, that part which is in the saturated zone of a defined aquifer. Sometimes called underflow.

Habitat guild—Groups of species that share common characteristics of microhabitat use and selection at various stages in their life histories.

High flow pulse—A short-duration, high flow within the stream channel that occurs during or immediately following storm events and serves to flush fine sediment deposits and waste products, restore normal water quality following prolonged low flows, and provide longitudinal connectivity for species movement along the river

Hydraulic control—A horizontal or vertical constriction in the channel, such as the crest of a riffle, which creates a backwater effect.

Hydrograph—A graph showing the variation in discharge over time.

Hydrologic regime—The distribution over time of water in a watershed, among precipitation, evaporation, soil moisture, groundwater storage, surface storage, and runoff.

Hyporheic zone—The interface between the stream bed and shallow ground water.

Index of biotic integrity—A numerical gauge of the biological health of stream fish communities based on various attributes of species richness, species composition, trophic relations, and fish abundance and condition.

Instantaneous flow—(1) Discharge that is measured at any instance in time. (2) Flow that is measured instantaneously and not averaged over longer time such as day or month.

Instream flow—The rate of flow in a natural stream channel at any time of year.

Instream flow requirement—(1) That amount of water flowing through a natural stream course that is needed to sustain, rehabilitate or restore the ecological functions of a stream in terms of hydrology, biology, geomorphology, connectivity and water quality at a particular level. (2) That amount of water flowing in a stream needed to sustain the protection of fish and wildlife habitat, migration, and propagation; outdoor recreation activities; navigation; hydropower generation; waste assimilation (water quality); and ecosystem maintenance, which includes recruitment of fresh water to the estuaries, riparian vegetation, floodplain wetlands, and maintenance of channel geomorphology. Instream flow requirements are typically recognized and administered under the authority of some type of legal means such as a water right, permit or operating agreement.

Instream Flow Incremental Methodology (IFIM)—Modular decision support system for assessing potential flow management schemes. It quantifies the relative amounts of total habitat available for selected aquatic species under proposed alternative flow regimes.

Instream use—Any use of water that does not require diversion or withdrawal from the natural watercourse, including in-place uses such as navigation and recreation.

Large woody debris—Any large piece of woody material that intrudes into the stream channel; often defined as having a diameter greater than 10cm and a length greater than 1m. Synonyms: Large organic debris, woody debris, log.

Macrohabitat—Abiotic habitat conditions in a segment of river controlling longitudinal distribution of aquatic organisms, usually describing channel morphology, flow, or chemical properties or characteristics with respect to suitability for use by organisms.

Main stem—The main channel of a river, as opposed to tributary streams, and oxbow lakes or floodplain sloughs.

Mesohabitat—A discrete area of stream exhibiting relatively similar characteristics of depth, velocity, slope, substrate, and cover, and variances thereof (e.g., pools with maximum depth <5 ft, high gradient riffles, side channel backwaters).

Microhabitat—Small localized areas within a broader habitat type used by organisms for specific purposes or events, typically described by a combination of depth, velocity, substrate, or cover.

Minimum flow—The lowest streamflow required to protect some specified aquatic function as established by agreement, rule, or permit.

Natural flow—The flow regime of a stream as it occurs under completely unregulated conditions; that is, a stream not subjected to regulation by reservoirs, diversions, or other human works.

Naturalized flow—Managed flows that are adjusted to mimic flows that would occur in the absence of regulation and extraction.

Normal flow—See base flow.

Open channel hydraulics—The analysis of water flow and associated materials in an open channel with a free water surface, as opposed to a tunnel or pipeline.

Overbank flow—An infrequent, high flow event that overtops the river banks, physically shapes the channel and floodplain, recharges ground water tables, delivers nutrients to riparian vegetation, and connects the channel with flood plain habitats that provide additional food for aquatic organisms.

PHABSIM—The Physical HABitat SIMulation system. A set of software and methods that allows the computation of a relation between stream flow and physical habitat for various life stages of an aquatic organism or a recreational activity.

$Q7_{10}$—The lowest continuous 7-day flow with a 10-year recurrence interval. Sometimes called $7Q_{10}$.

Reach—A comparatively short length of a stream, channel, or shore. One or more reaches compose a segment.

Riffle— A relatively shallow reach of stream in which the water flows swiftly and the water surface is broken into waves by obstructions that are completely or partially submerged.

Riparian/riparian zone—Pertaining to anything connected with or adjacent to the bank of a stream or other body of water. The transitional zone or area between a body of water and the adjacent upland identified by soil characteristics and distinctive vegetation that requires an excess of water, including wetlands, marshes, and floodplains that support riparian vegetation.

Riparian vegetation—Vegetation that is dependent upon an excess of moisture during a portion of the growing season on a site that is perceptively more moist than the surrounding area.

Sediment—Solid material, both mineral and organic, that is in suspension in the current or deposited on the streambed.

Sediment load—A general term that refers to material in suspension and/or in transport. It is not synonymous with either discharge or concentration. (See Bedload).

Segment—A relatively long (e.g., hundreds of channel widths) section of a river, exhibiting relatively homogeneous conditions of hydrology, channel geomorphology, and pattern. **Stream**—A natural watercourse of any size containing flowing water, at least part of the year, supporting a community of plants and animals within the stream channel and the riparian vegetative zone.

Streambed—The bottom of the stream channel; may be wet or dry.

Subsistence flow—The minimum streamflow needed during critical drought periods to maintain tolerable water quality conditions and provide minimal aquatic habitat space for the survival of aquatic organisms.

Suspended sediment—Particles that are suspended in the moving water column for long distances downstream. Much of this material settles out when water movement slows or ceases.

Time-series analysis—Analysis of the pattern (frequency, duration, magnitude, and time) of time-varying events. These events may be discharge, habitat areas, stream temperature, population factors, economic indicators, power generation, and so forth.

Tributary—A stream feeding, joining, or flowing into a larger stream (at any point along its course or into a lake). Synonyms: feeder stream, side stream.

Turbidity—A measure of the extent to which light passing through water is reduced due to suspended materials.

Water allocation—Determining the quantity of water from a given source that can or should be ascribed to various instream or out-of-stream uses. May be referred to as water reservation in some settings.

Water resources—The supply of ground water and surface water in a given area.

Water right—A legally protected right to use surface or groundwater for a specified purpose (such as crop irrigation or water supply), in a given manner (such as diversion or storage), and usually within limits of a given period of time (such as June through August). While such rights may include the use of a body of water for navigation, fishing, hunting, and other recreational purposes, the term is usually applied to the right to divert or store water for some out-of-stream purpose or use.

Watershed— See Drainage area.

Wetted perimeter—The distance along the stream bottom from the wetted edge on one side to the wetted edge on the other measured at a given discharge.

SOURCE: Adapted from IFC, 2002.

Appendix B

WATER SCIENCE AND TECHNOLOGY BOARD

R. RHODES TRUSSELL, *Chair*, Trussell Technologies, Inc., Pasadena, California
RICHARD G. LUTHY[*], *Chair*, Stanford University, Stanford, California
JOAN B. ROSE[*], *Vice Chair*, Michigan State University, East Lansing
RICHELLE M. ALLEN-KING[*], Washington State University, Pullman
MARY JO BAEDECKER, U.S. Geological Survey (retired), Reston, Virginia
GREGORY B. BAECHER, University of Maryland, College Park
KENNETH R. BRADBURY[*], Wisconsin Geological and Natural History Survey, Madison
JAMES CROOK[*], Water Reuse Consultant, Norwell, Massachusetts
JOAN G. EHRENFELD, Rutgers University, New Brunswick, New Jersey
DARA ENTEKHABI, Massachusetts Institute of Technology, Cambridge
EFI FOUFOULA-GEORGIOU[*], University of Minnesota, Minneapolis
GERALD GALLOWAY, Titan Corporation, Reston, Virginia
PETER GLEICK, Pacific Institute for Studies in Development, Environment, and Security, Oakland, California
CHARLES N. HAAS, Drexel University, Philadelphia, Pennsylvania
KAI N. LEE, Williams College, Williamstown, Massachusetts
JOHN LETEY, JR. [*], University of California, Riverside
CHRISTINE L. MOE, Emory University, Atlanta, Georgia
ROBERT PERCIASEPE, National Audubon Society, Washington, D.C.
JERALD L. SCHNOOR, University of Iowa, Iowa City
LEONARD SHABMAN, Resources for the Future, Washington, D.C.
KARL K. TUREKIAN, Yale University, New Haven, Connecticut
HAME M. WATT, Independent Consultant, Washington, D.C.
CLAIRE WELTY, University of Maryland, Baltimore County, Baltimore
JAMES L. WESCOAT, JR., University of Illinois at Urbana-Champaign

Staff

STEPHEN D. PARKER, Director
LAURA J. EHLERS, Senior Program Officer
JEFFREY W. JACOBS, Senior Program Officer
WILLIAM S. LOGAN, Senior Program Officer
LAUREN E. ALEXANDER, Program Officer
MARK C. GIBSON, Program Officer
STEPHANIE E. JOHNSON, Program Officer
M. JEANNE AQUILINO, Financial and Administrative Associate
ELLEN A. DE GUZMAN, Research Associate
PATRICIA JONES KERSHAW, Senior Program Associate
ANITA A. HALL, Program Associate
DOROTHY K. WEIR, Senior Program Assistant

* Terms expired June 30, 2004.

Appendix C

Biographical Sketches for Committee on Review of Methods for Establishing Instream Flows for Texas Rivers

Gail E. Mallard, *Chair*, is with the U.S. Geological Survey where she serves as a Senior Advisor to the Associate Director for Water. She is the co-Chair of the National Water Quality Monitoring Council. She has over 15 years of experience in planning and managing water resources programs, including planning for the USGS National Water Quality Assessment Program, the USGS National Research Program, and the USGS Toxic Substances Hydrology Program. Within the USGS, she has also provided advice on workforce planning issues and technical support for water resources programs. She served as Chair of the Freshwater Work Group and member of the design Committee for the Heinz Center report, "The State of the Nation's Ecosystems: Measuring the Lands, waters, and Living Resources of the United States". Her technical interest and expertise is water quality and environmental monitoring. She received a Ph.D. from the Ohio State University in microbiology in 1975.

Kenneth L. Dickson is the Regents Professor of Biological Sciences at the University of North Texas. Dr. Dickson created the Elm Fork Education Center in 1998 and serves as its Director. He is past Director of the Institute of Applied Sciences at UNT. Dr. Dickson conducts research in applied problem solving in aquatic biology, development of methods to evaluate the fate and effects of chemicals in the aquatic environment, hazard assessment, biomonitoring, aquatic toxicology, limnology of reservoirs, restoration and recovery of damaged ecosystems, environmental education, and applications of remote sensing and GIS to environmental impact assessment. Dr. Dickson earned a BS in education (1966) and an MS in biology (1968) from North Texas State University, and a Ph.D. in aquatic biology from Virginia Polytechnic Institute and State University (1971).

Thomas B. Hardy is the Associate Director for the Utah Water Research Laboratory and also serves as a professor of Biological and Irrigation Engineering at Utah State University. Dr. Hardy has worked for advancements in hydraulic simulation techniques for use in stream habitat modeling and optimization techniques in water resource allocation studies involving in-stream flow determinations for fisheries, application of multispectral remote sensing techniques for the classification and quantification of stream/riparian ecosystem elements for GIS applications, development of computer simulation models and software interfaces for use with assessment tools. Dr. Hardy earned BS degrees in both education (1977) and biology (1978) from the University of Nevada, and an MS degree in aquatic biology (1982) also from the University of Nevada. Dr. Hardy received his Ph.D. from Utah State University in civil engineering in 1988.

Clark Hubbs is the Regents Professor of zoology, Emeritus, of the University of Texas at Austin. Dr. Hubbs studies how fish relate to their environment and how anthropogenic changes impact their survival. He investigates the causes and cures of endangered species status. His studies involve geographic variation in life history traits and interactions between a gynogenetic sexual parasite and its male host species and the reasons for the differences between spring and stream aquatic biota. Dr. Hubbs received an AB in zoology from the University of Michigan in 1942, and a Ph.D. from Stanford University in 1951.

David R. Maidment is the Engineering Foundation Professor of Civil Engineering and Director of the Center for Research in Water Resources at the University of Texas at Austin. His current research involves the application of geographic information systems to floodplain mapping, water-quality modeling, water resources assessment, hydrologic simulation, surface water–groundwater interaction, and global hydrology. In 2003, Dr. Maidment received the Lifetime Achievement Award of the Environmental Systems Research Institute (ESRI) and was named a National Associate of the National Academies. Dr. Maidment has served the NRC as chair of both the Committee on Water Resources Research and the Committee on Review of the USGS National Streamflow Information Program. From 1992 to 1995 he was Editor of the Journal of Hydrology, and he is currently an associate editor of that journal. He received his B.S. degree in Agricultural Engineering from the University of Canterbury, Christchurch, New Zealand, and his M.S. and Ph.D. degrees in Civil Engineering from the University of Illinois at Urbana-Champaign.

James Martin is the executive director of Western Resource Advocates (WRA), a non-profit environmental law and policy organization dedicated to restoring and protecting the natural environment of the Interior American West. Before joining WRA, he served as director of the Natural Resources Law Center at the University of Colorado, School of Law, where he conducted research on a wide range of public lands and resources issues and taught advanced natural resources law seminars on land use planning and energy law. Mr. Martin also previously served as a senior attorney at Environmental Defense where he worked on air quality, energy, endangered species, and water resources issues. From 1986 to 1992, he served Congressman and then Senator Tim Wirth as counsel for energy, environment and natural resources and as the senator's state director. He has a bachelor's degree in biology from Knox College (1973) and a J.D. degree (1981) with a certificate in environmental and natural resources law from Northwestern School of Law of Lewis and Clark College in Portland, Oregon.

Patricia F. McDowell is Professor of geography and Professor of environmental studies at the University of Oregon. She teaches courses in fluvial geomorphology, watershed science and policy, and soils geography. Her research focuses on response of river systems to human impacts and environmental change. At the University of Oregon, she served as Associate Vice President for Research from 1990 to 1993 and as Chair of the Department of Geography from 1993 to 1996. She has served the NRC as a member on the Committee on Research Priorities in Geography at the USGS. Dr. McDowell earned a BA (1971) and MA (1977) from the Illinois Institute of Technology, and a Ph.D. (1980) from the University of Wisconsin.

Brian D. Richter is the director of The Nature Conservancy's Sustainable Waters Program, an international effort to protect freshwater systems. Brian Richter has been involved in river conservation for more than 20 years. His current responsibilities focus on the global challenges of meeting human needs for water while keeping river ecosystems healthy. He works with public agencies, academic institutions, and other private organizations involved in river conservation, and he leads a staff that includes hydrologists, aquatic ecologists, policy specialists, educators and communicators. He has published numerous scientific papers on the importance of ecologically sustainable water management in international science journals. He has also co-authored a new book with Sandra Postel entitled *"Rivers for Life: Managing Water for People and Nature,"* published by Island Press in summer 2003.

Gregory V. Wilkerson is an Assistant Professor in the Department of Civil and Architectural Engineering at the University of Wyoming. Dr. Wilkerson's research interests include research and development of solutions to water resource problems, multi-disciplinary approaches to stream restoration, river mechanics, sedimentation and erosion, environmental hydraulics, engineering hydrology, and statistics. His current research involves developing improved methods for physical modeling of rivers and developing a GIS program for predicting the impact of increased water discharges, a by-product of coal-bed methane production, into natural rivers. Dr. Wilkerson is currently a P.I. with the NSF Science and Technology Center, National Center for Earth-Surface Dynamics. Dr. Wilkerson earned a BS degree in civil engineering from the Georgia Institute of Technology in 1989, and an MS (1995) and a Ph.D. (1999) both in civil engineering from Colorado State University.

Kirk O. Winemiller is a Professor in the Department of Wildlife and Fisheries Sciences at Texas A&M University. Dr. Winemiller earned a BA (1978) and an MS (1981) in zoology from Miami University in Oxford, Ohio, and a Ph.D. (1987) from the University of Texas, Austin, in 1987. Prior to joining the faculty at Texas A&M, Dr. Winemiller was Research Associate in the Environmental Sciences Division of the Oak Ridge National Laboratory where he worked on models of fish population dynamics as a member of the CompMech team. He is former Associate Editor for the Journal of Fish Biology and Transactions of the American Fisheries Society, and is currently Associate Editor for Ecology and Ecological Monographs. Dr. Winemiller's lab conducts field research on the ecology and management of fishes and macroinvertebrates in streams, rivers, and estuaries in Texas, including studies designed to develop and test ecological assessment tools. He also has over 20 years of experience investigating fish ecology and ecosystem dynamics in tropical rivers and estuaries.

David A. Woolhiser (NAE) received his Ph.D. in civil engineering from the University of Wisconsin in 1962. He retired from the USDA Agricultural Research Service in 1991 after a 30 year career and is currently a hydrologist in Fort Collins, Colorado. Since retirement he has served as Faculty Affiliate in civil engineering at Colorado State University. Dr. Woolhiser is known for his work on the hydrology and hydrometeorology of arid and semiarid rangelands, simulation of hydrologic systems, numerical modeling of surface runoff, erosion and chemical transport, and probabilistic models of rainfall and runoff. He was elected as a member to the National Academy of Engineering in 1990 for advancing the use of mathematical and statistical techniques to rationalize the description of hydrologic phenomena. Dr. Woolhiser has served the NRC on several committees, including the Committee on Water Resources Research, the Special Fields and Interdisciplinary Engineering Peer Committee, and the Steering Committee on Climate Change and Water Resources Management.

STAFF

Lauren E. Alexander is a program officer with the National Research Council's Water Science and Technology Board. Her research interests include hydro-geomorphic processes and plant diversity in forested wetlands, and she has studied forested wetlands in different coastal plain systems in the United States. Dr. Alexander received her B.S. in applied mathematics and her Masters of Planning in environmental planning from the University of Virginia, and her Ph.D. in landscape ecology from Harvard University. She joined the NRC in 2002.

Dorothy K. Weir is a senior program assistant with the Water Science and Technology Board. She received a BS in biology from Rhodes College in Memphis, Tennessee and is currently pursuing an MS degree in environmental science and policy from Johns Hopkins University. Ms. Weir joined the NRC in 2003.